CONTENTS

S. HRG. 107–1103

TRANSITION TO DIGITAL TELEVISION

HEARING

BEFORE THE

COMMITTEE ON COMMERCE, SCIENCE, AND TRANSPORTATION UNITED STATES SENATE

ONE HUNDRED SEVENTH CONGRESS

FIRST SESSION

MARCH 1, 2001

Printed for the use of the Committee on Commerce, Science, and Transportation

U.S. GOVERNMENT PRINTING OFFICE

87–414 PDF WASHINGTON : 2005

For sale by the Superintendent of Documents, U.S. Government Printing Office
Internet: bookstore.gpo.gov Phone: toll free (866) 512–1800; DC area (202) 512–1800
Fax: (202) 512–2250 Mail: Stop SSOP, Washington, DC 20402–0001

COMMITTEE ON COMMERCE, SCIENCE, AND TRANSPORTATION

ONE HUNDRED SEVENTH CONGRESS

FIRST SESSION

JOHN McCAIN, Arizona, *Chairman*

TED STEVENS, Alaska
CONRAD BURNS, Montana
TRENT LOTT, Mississippi
KAY BAILEY HUTCHISON, Texas
OLYMPIA J. SNOWE, Maine
SAM BROWNBACK, Kansas
GORDON SMITH, Oregon
PETER G. FITZGERALD, Illinois
JOHN ENSIGN, Nevada
GEORGE ALLEN, Virginia

ERNEST F. HOLLINGS, South Carolina
DANIEL K. INOUYE, Hawaii
JOHN D. ROCKEFELLER IV, West Virginia
JOHN F. KERRY, Massachusetts
JOHN B. BREAUX, Louisiana
BYRON L. DORGAN, North Dakota
RON WYDEN, Oregon
MAX CLELAND, Georgia
BARBARA BOXER, California
JOHN EDWARDS, North Carolina
JEAN CARNAHAN, Missouri

MARK BUSE, *Republican Staff Director*
MARTHA P. ALLBRIGHT, *Republican General Counsel*
KEVIN D. KAYES, *Democratic Staff Director*
MOSES BOYD, *Democratic Chief Counsel*

TRANSITION TO DIGITAL TELEVISION

THURSDAY, MARCH 1, 2001

U.S. SENATE,
COMMITTEE ON COMMERCE, SCIENCE, AND TRANSPORTATION,
Washington, D.C.

The Committee met, pursuant to notice, at 9:35 a.m. in room SR–253, Russell Senate Office Building, Hon. John McCain, Chairman of the Committee, presiding.

OPENING STATEMENT OF HON. JOHN McCAIN, U.S. SENATOR FROM ARIZONA

The CHAIRMAN. Good morning. In 1997 the Congress embarked on a path that we hoped would lead to a revolution for the American consumer: digital television. Digital television technology—especially high definition television—we were told has the potential to provide sharp movie-quality pictures as well as CD-quality sound to the U.S. consumer.

The Congress took some extraordinary steps in pursuit of the DTV goal, giving to the broadcast industry a huge amount of spectrum and engaging in what the *Wall Street Journal* then described as a "planned multi-billion dollar handout for wealthy TV-station owners." Although other industries must buy their spectrum in competitive auctions, here, the government decided to give away the spectrum needed to provide free TV broadcasts in high definition television. Moreover, the government decided that the broadcasters could keep their old analog spectrum—a gift from the past—until 2006, or until 85 percent of American homes had digital television.

Considering it took approximately 20 years for color TV and 16 years for VCRs to reach that level of market penetration, that was quite a gift.

And now where are we? The situation is a mess—characterized more by finger pointing than progress.

Manufacturers blame the dearth of digital programming for low consumer demand for HDTV, and the cable companies blame the manufacturers for delays in agreeing to interoperability standards. And the broadcasters, well, the broadcasters blame the FCC, local zoning boards, standards disputes, equipment manufacturers, content providers and Congress—in short, everybody but themselves.

And then there is still the question of standards and whether we actually are moving to digital TV of HDTV.

An attempt to assign blame for this situation is futile.

The interested parties all disagree about whether there is a problem, the extent of any problem, and who is to blame. But this much is clear: By 2006, this country will have neither the transmission facilities, nor the digital content, nor the reception equipment needed to ensure that 85 percent of the population will be able to received digital television as their exclusive source of television.

Congress, having given away billions of dollars in public assets to ensure a smooth and rapid transition to a competitive DTV marketplace, will be held accountable by the public if that transition becomes even slower, more costly to the taxpayer, or anticompetitive.

In short, no matter who is to blame for the existing problems, the Congress that devoted public assets to the DTV transition will be held accountable for finding solutions.

But many of the solutions being proposed today presume that problems created by a failed attempt at centralized planning can be solved by more attempts at centralized planning. Worse yet, some of these proposed solutions would seem only to shift additional costs of the DTV transition onto the backs of taxpayers who have already devoted their valuable spectrum to the transition, or onto the backs of competing industries that pay for their spectrum.

For example, some propose mandating that all television sets sold in America be made digitally compatible—even though this would substantially raise the cost of a new television set. Others propose that broadcasters should be allowed to benefit from a slow DTV transition by gaining indefinite use of free spectrum that could be used to multicast standard definition signals or to distort competition in the wireless communications markets by competing against companies and technologies that had to pay for the spectrum they use. We must examine all these issues here today.

Political columnist William Safire noted in the *New York Times*, quote: In terms of ripping off the taxpayers with not a peep from the media, nothing compares with the broadcasters' lobby. This phalanx of freeloaders has stolen the free use of great chunks of the most valuable natural resource of the information age: the digital television spectrum owned by the American people. When private money is on the line, private companies move fast; but when public assets go to private pockets, at no interest, private companies sit tight."

We are here at this hearing to give the broadcast industry a chance to show its commitment to sound public policy—to show that it will do more than just sit tight.

I thank today's witnesses for joining us and look forward to their testimony.

Senator Stevens, do you have an opening comment?

Senator STEVENS. Senator Burns was here first, Mr. Chairman.

The CHAIRMAN. I am sorry. We usually go by seniority as to when the hearing starts, but if you would like, Senator Burns, you are recognized.

Senator BURNS. Go ahead, Senator.

STATEMENT OF HON. TED STEVENS, U.S. SENATOR FROM ALASKA

Senator STEVENS. Mr. Chairman, I think we should listen to the industry today. What we planned, obviously you are right, was not on track and the market has not responded to the timeframe that we originally envisioned for this transition.

There are a lot of unique problems out there in converting to digital. In my State, for instance, one of the great problems is the terrain, and we find that instead of one tower in Anchorage, we are going to have to have two towers. If we put it high enough for one tower, it would interfere with the FAA restrictions to protect aircraft.

We are going to have to have probably some specific legislation dealing with rural areas such as ours, but beyond that, we have got the problem of the public broadcasters. We did put up money to assist the small broadcasters to make the transition, but it was subject to authorization. Because no authorization bill has passed, the funds are not available for use. By definition, public broadcasters are going to be out probably 2 or 3 years beyond the deadline in existing law before they can make the conversions.

I do think there is some need for extension of the deadlines in the existing law. I hear what you say and I respectfully disagree with some of it. I think we have got to listen to the market and see what the market is going to do with regard to development of demand for this new digital service. These people just cannot afford to convert ahead of the public and yet, the public will not convert until they start the mechanism, as you say, of conversion.

So, it is time for us to listen, I think, and see what the respective portions of the industry advise us to do. Clearly, I think this Congress is going to have to do something, or else the deadlines in the existing law will come, and there is going to be real chaos out there in about 1 year. I hope this is just the first of a series of hearings to deal with this issue, because I think we are going to hear some rather divergent views here today. As I said, I think we should listen to them before we make our final judgment of what to do. Thank you.

The CHAIRMAN. Thank you, Senator Stevens.

Senator Wyden.

STATEMENT OF HON. RON WYDEN, U.S. SENATOR FROM OREGON

Senator WYDEN. Thank you, Mr. Chairman. I very much agree with what both you and Senator Stevens have said and it just seems to me we have got a classic chicken and egg situation here. Nobody wants to broadcast digital signals if consumers do not have TVs to receive them, and consumers do not want to buy the TVs to receive digital signals if nobody is broadcasting them. And you characterized it, Mr. Chairman, I think correctly, we have just got a lot of finger-pointing.

The one area that I would like to explore is whether it might be possible to create some incentives on both sides of this debate, consumers and broadcasters, to encourage people to move more quickly. I am reluctant to move the dates back, because I think everything would fall apart if that were the case, but as both you and Senator Stevens have said, we do need to listen today and look for some creative ways to move forward, and I appreciate you holding the hearing.

The CHAIRMAN. Thank you.

Senator Burns.

STATEMENT OF HON. CONRAD BURNS, U.S. SENATOR FROM MONTANA

Senator BURNS. Mr. President, I would ask unanimous consent that I put my statement in the record.

The CHAIRMAN. Without objection.

Senator BURNS. Just with a comment, though, that I looked back and looked back over previous statements regarding this, and I think we took sort of a cautious approach to conversion. It seems strange to me that we go around the world promoting a market driven economy, and we said at the time that the conversion will be driven more by market than it will be by government mandates, and on how we make our investments in the broadcast industry and those things.

I agree with you though, there are some problems out there engineering wise, and like in my State of Montana, and my State sort of mirrors, although not on near the scale that Alaska does, but we are confronted by some of the same problems.

There are some components, and I think I would agree with both of you that we had better sit and listen today to the experts, and do some of our own individual investigation on what is truly going on in the market and how the progress is being made before we make any decisions that we would probably not like in 2 or 3 years. But we are coming down to a deadline where we are going to have to make some decisions.

I thank you for holding this hearing today. I think it is very very important.

[The prepared statement of Senator Burns follows:]

PREPARED STATEMENT OF HON. CONRAD BURNS, U.S. SENATOR FROM MONTANA

I thank the Chairman for calling the hearing today, as it concerns a topic that is both vitally important and often misunderstood—the transition to digital television. I think it's very appropriate that this hearing was elevated to the full committee level and I thank the Chairman for doing so.

I share your interest in ensuring that the transition move forward as expeditiously as possible. The American people have a significant investment in this transition. We therefore owe it to the American people to make sure that this valuable resource is put to its best use. This includes ensuring that broadcasters' analog spectrum is returned at the earliest possible date so that spectrum can be re-allocated for other uses, such as next-generation mobile services.

I should note that most broadcasters are ready and willing to move forward and keep the promises they made to Congress over the last several years. Indeed, many broadcasters have already invested many millions of dollars in creating the infrastructure necessary for a rapid transition to digital programming.

I've always been skeptical of government mandates, just as I felt that hard-and-fast digital buildout requirements were more a product of budget politics than engineering reality. Rather, the debate should focus on the progress the broadcasters are making in their conversion to this exciting new technology and how it can be made available to the public as quickly as possible.

I look forward to hearing from today's witnesses on the state of the transition. We need to analyze each of the three components to the transition, and assess where we are in the process.

The first component is the broadcast stations' build-out requirements. The FCC has established firm build-out deadlines. The Committee needs to know where we stand on those deadlines. Are there any regulatory obstacles blocking our path to an expeditious build out?

The second component relates to programming. Are program producers creating enough digital programming? Is that programming being filmed in analog format, and then converted to digital? Or is the programming originally filmed in digital format?

The last of the three components relates to the American consumer. Like the stations and the program producers, consumers have to convert to digital as well. And in order for them to convert, they must have access to a robust DTV product market that offers consumers competitive choices and prices.

On this point, I am concerned that the development of a robust DTV product is being slowed by the delay in industry negotiations over copy protection. Specifically, I understand that the negotiations are deadlocked over the question of whether DTV devices should be designed in such a that they protect against retransmission of local broadcast programming over the Internet.

In my view, industry should agree to provide the same level of copy protection to broadcast programming that it provides come to non-broadcast programming. Through the recent satellite television legislation, Congress re-affirmed its commitment to free, over-the-air television. I urge industry negotiators to renew their efforts in such a way that our commitment to free, over-the-air television is embodied in copy protection negotiations.

In doing so, industry can do its part to encourage innovation in the DTV product market, which in turn will speed the transition to digital.

Thank you, Mr. Chairman. I look forward to the testimony of today's witnesses.

The CHAIRMAN. Thank you very much, Senator Burns.

Our first panelist, Mr. Jeff Sagansky, is the President-CEO of Paxson Communications; Mr. Ben Tucker is the Executive Vice President for Broadcast Operations of Fisher Broadcasting; and Mr. Michael Willner is President and CEO of Insight Communications.

I would like to welcome our first panel of witnesses, and Mr. Sagansky, we will begin with you.

STATEMENT OF JEFF SAGANSKY, PRESIDENT AND CHIEF EXECUTIVE OFFICER, PAXSON COMMUNICATIONS CORPORATION

Mr. SAGANSKY. Thank you, Mr. Chairman and distinguished members of the Committee for providing me with the opportunity to appear before your panel today to discuss the digital television transition. My name is Jeff Sagansky and I am President and CEO of Paxson Communications.

Senator BURNS. A little closer with the microphone, thank you.

Mr. SAGANSKY. I am CEO of Paxson Communications Corporation, the largest broadcast television group owner in the United States, 65 stations strong, and the creator of PAX-TV network, which was launched in August 1998 and now reaches 81 percent of all American homes.

Throughout my career, I have programmed various broadcast networks that became No. 1 with family programming. At NBC in the 1980s, I programmed "Cosby" and "Family Ties." At CBS in the 1990s, "Touched by an Angel" and "Dr. Quinn Medicine Woman." This type of programming is no longer favored by the broadcast networks, so at PAX-TV we set out to create a network that was dedicated to family-oriented programming, with an emphasis on positive values and strong role models.

When we launched two-and-a-half years ago, we repeatedly heard from media pundits—"no sex, no violence, no ratings." Yet, here we are today, turning a profit and proving that the public is looking for—and advertisers will support—family television.

We would not exist today at all were it not for the ability of our owned stations and our affiliated stations to obtain cable and satellite carriage for their signals under the 1992 Cable Act and the SHVIA Act of 1999.

Our future as an emerging network is now tied directly to the success of digital television transition in this country and to the continued full cable and satellite carriage of our stations' programming. That digital future is a potentially great one. Some networks see the future of high definition TV showing sporting events and movies. We at PAX have always maintained that the highest and best use of our digital spectrum is multiple channels of high quality information, lifestyle news and entertainment centered around the family; enabling families to be more effective in their homes and in their communities.

And yet, the digital transition that is our future is in very serious trouble. Within 14 months our 65 stations must be broadcasting digitally. However, at this hour only a third of our 65 stations have even received DTV construction permits from the FCC.

Last year in the United States there were 33 million analog TV sets sold, compared to only 26 thousand digital tuners.

We need a digital All-Channel Receiver Act that would enable all television sets sold to the American public to be capable of receiving both analog and digital signals. Consumers have a right, when they buy a TV set, to be assured that it will not shortly become obsolete.

There are also copyright issues and cable-television interoperability issues that must be promptly resolved once and for all by the FCC. After 4 years, it is obvious that the marketplace is not establishing these standards. The FCC needs to deal with these issues now.

Undoubtedly, the most important single issue for PAX-TV in terms of a successful DTV transition to cable is cable and satellite carriage of all 6 Mhz of our stations' digital signals.

Like us, many other broadcasters, including public television stations, believe that the capability to multicast several programming services is the key to their use of the digital spectrum. We do not feel that our digital future is in devoting our entire digital capacity to a single stream of programming, nor in using our digital capacity for ancillary uses such as datacasting. But we need the assurance that our multiple free, over-the-air programming services will be received by 70 percent of the homes in this country that are served by cable and satellite. They are the gatekeeper.

Our concern is that a divided FCC last month adopted rules that will not only hurt the DTV transition, but undermine PAX-TV's efforts to multicast its free, over-the-air programming services. First, the FCC said that television stations cannot request cable carriage of their digital signal until they turn in their analog channels, which will be years away. We think this is a bad decision. This discourages broadcasters from building their digital stations and consumers from buying DTV sets, and it severely damages the overall chances for economic viability of digital television.

The FCC also decided to permit cable operators to carry only one of a station's multiple channels of free, over-the-air programming rather than requiring cable systems to carry all such free programming. This decision was also wrong. It is contrary to the congressional intent evidence in 1992 when you adopted the must-carry rules. Anyone reading the recent FCC decision will recognize that the Commissioners were clearly uneasy with their decision and, in fact, were reaching out to Congress for guidance on this issue. PAX-TV urges this Committee to take the opportunity to reaffirm the congressional commitment to full digital must-carry and to the preservation of free, local television by endorsing multi-channel digital must-carry for all free, over-the-air programming services.

If content is going to be one of the key drivers to the DTV transition, then give the consumer access to all the free content that we, the local broadcasters, have the ability to air.

Thank you.

[The prepared statement of Mr. Sagansky follows:]

PREPARED STATEMENT OF JEFF SAGANSKY, PRESIDENT AND CHIEF EXECUTIVE OFFICER, PAXSON COMMUNICATIONS CORPORATION

Thank you Mr. Chairman, Mr. Co-Chairman and distinguished Members of the Committee for providing me with the opportunity to appear before your panel today to discuss the Digital Television Transition. My name is Jeff Sagansky and I am President and CEO of Paxson Communications Corporation, the largest television group owner in the United States, 65 stations strong, and the creator of the PAX-TV network, which was launched August, 1998 and now reaches 81 percent of all American Homes.

Throughout my career I have programmed various broadcast networks that became No. 1 with family programming. At NBC in the 1980s, I programmed "Cosby" and "Family Ties." At CBS in the 1990s, I developed "Touched By An Angel" and "Dr. Quinn Medicine Woman." This type of programming is no longer favored by the broadcast networks, so at PAX-TV we set out to create a network that was dedicated to family oriented programming with an emphasis on positive values and strong role models.

When we launched 2½ years ago, we repeatedly heard from media pundits—no sex, no violence, no ratings. And yet here we are today, turning a profit and proving that the public is looking for—and advertisers will support—family values television.

We would not exist today were it not for the ability of our owned and affiliate stations to obtain cable and satellite carriage for their signals under the 1992 Cable Act and the SHVIA Act of 1999.

Our future as an emerging network is now tied to the success of the digital television transition in this country and to the continued full cable and satellite carriage of our stations' programming. That digital future is a potentially great one. Some networks see a future of high definition TV showing sporting events and movies. We at PAX have always maintained that the highest and best use of our digital spectrum is multiple channels of high quality information, lifestyle news and entertainment centered on the family; enabling families to be more effective within their homes and communities.

And yet, the digital transition that is our future is in very serious trouble. Within 14 months, our 65 television stations must be broadcasting digitally. However, at this hour only one-third of our stations have even received DTV construction permits from the FCC.

Last year in the United States there were 33 million analog TV sets sold compared to only 26 thousand digital tuners.

We need a Digital All-Channel Receiver Act that would require that all television sets sold to the American public be capable of receiving both analog and digital signals. Consumers have a right when they buy a TV set to be assured that it will not shortly become obsolete.

There are also copyright issues and cable-television inter-operability issues that must be promptly resolved once and for all by the FCC. After 4 years, it is obvious the marketplace is not establishing these standards. The FCC needs to deal with these issues now.

Undoubtedly, the most important issue for PAX-TV in terms of a successful DTV transition is cable and satellite carriage of all 6 Mhz of our stations' digital signals.

Like us, many other broadcasters including public television stations believe that the capability to multicast several programming services is the key to their use of the digital spectrum. We don't feel our digital future is in devoting our entire digital capacity to a single stream of programming nor in using digital capacity for ancillary uses such as datacasting. But we need the assurance that our multiple free, over-the-air programming services will be received by the 70 percent of the homes of this country that are served by cable and satellite.

Our concern is that a divided FCC last month adopted rules that not only will hurt the DTV transition but undermine PAX-TV's efforts to multicast its free, over-the-air family program services. First, the FCC said that television stations cannot request cable carriage of their digital signal until they turn in their analog channels which will be years away. We think this is a bad decision that discourages broadcasters from building their digital stations and consumers from buying DTV sets and severely damages the chances for economic viability of digital television.

The FCC also decided to permit cable operators to carry only one of a station's multiple channels of free, over-the-air programming rather than requiring cable systems to carry all such free programming. This decision was also wrong. It is contrary to the Congressional intent evidenced in 1992 when you adopted the must-carry rules. Anyone reading the recent FCC decision will recognize that the Commissioners were clearly uneasy with their decision and, in fact, were reaching out to Congress for guidance on this issue. PAX-TV urges this Committee to take the opportunity to reaffirm the Congressional commitment to full digital must-carry and to the preservation of free, local television by endorsing multi-channel digital must-carry of all free, over-the-air programming services.

If content drives the DTV transition, then give the consumer access to all the free multichannel content that we, the local broadcasters, have the ability to air.

The CHAIRMAN. Thank you very much.
Mr. Tucker.

STATEMENT OF BEN TUCKER, EXECUTIVE VICE PRESIDENT FOR BROADCAST OPERATIONS, FISHER BROADCASTING, INC., AND TELEVISION BOARD CHAIRMAN, NATIONAL ASSOCIATION OF BROADCASTERS

Mr. TUCKER. Thank you, Mr. Chairman, for the opportunity to appear before your Committee today to discuss the transition to digital television. My name is Ben Tucker. I am the Executive Vice President of Broadcast Operations for Fisher Broadcasting. We own 12 television stations in the States of Washington, Oregon, Idaho and Georgia, in both large and small markets. I am also the National Association of Broadcasters Television Board Chairman. I am pleased to represent the broadcasting industry today to provide an interim status report on the transition to digital television.

I would like to point out that the DTV transition is all about consumers. We want to provide the American public with all the benefits of digital television. Right now, the broadcasting industry is working hard in achieving this goal. We have 183 digital stations on the air as of this morning. These stations reach over two-thirds of the TV households across the Nation. 71 of these stations currently on the air—almost 40 percent—are ahead of their required build-out schedule.

Other stations may not be so lucky. In 14 months from now, all commercial television stations are supposed to be operating with a digital signal. This means there are approximately 1,100 stations yet to be on the air. We have seen some problems crop up with the DTV stations that have made it on the air. These involve delays in equipment delivery, local zoning board delays, delays from the FCC, and a shortage of tower crews. These issues will continue to impact local stations as they work to get on the air.

In my company, Fisher Broadcasting, we currently have 2 DTV stations on the air, KATU in Portland, Oregon, and KOMO in Seattle, Washington. And after the earthquake yesterday, I would like to say that we are still on the air. Both are ABC affiliates. We have ordered DTV equipment for all of our stations. I would like to highlight the fact that KOMO in Seattle is one of only three stations in the Nation that broadcasts its local news in high definition. As you can see, Fisher Broadcasting, like other broadcasters, is committed to the DTV transition as quickly as possible.

However, to get the job done, the other parties need to come to the table. There are four unlikely partners linked in this transition. The first three are broadcasters, consumer electronics manufacturers who make the televisions, and cable, the gatekeepers. The congressionally-appointed traffic cop, the FCC, is the fourth partner.

Our goal is to reach our viewers as quickly as possible. It is not, and I want to emphasize not, in the best interest of the broadcasting community to delay the DTV transition. However, for your constituents—our viewers—to enjoy the wonderful benefits of digital television, we need three things.

Number one, we do need must-carry. During the transition, digital cable systems must be required to carry both the analog and digital channels of local broadcasters. With 70 percent of the American public getting their broadcast channels through cable, cable cannot be allowed to act as the digital gatekeeper.

As the general counsel of the Nation's largest cable operator told the FCC last year, once the upgrade to digital cable, they will be crying for content. After the transition, these same cable operators must-carry all of the preprogramming offered by digital broadcasters. America's consumers deserve to have all of the benefits of free over-the-air digital television.

Number two, DTV sets. Set manufacturers must include DTV tuners in all new sets. Every set sold needs to have both an analog and digital tuner. In 1962, Congress mandated the UHF tuners in every set. The digital tuners should be mandated in every set in 2002.

Number three, DTV-cable interoperability. Senators, it may surprise you, even if you bought a digital television set with a DTV tuner, you couldn't connect it to a digital cable set-top box. Why? Because the cable consumer electronics industries won't agree on a standard. Fixing this is the only way that the cable viewing public can see free over-the-air broadcast digital signals that will drive this transition to completion.

These are the hurdles that must be cleared. If these three things happen, consumer exposure and the acceptance of digital television will skyrocket, digital prices will fall, and the American consumer will enjoy the benefits of digital television.

What does all of this mean? The broadcasting industry is working toward the target set by Congress to make this DTV transition. We cannot make it work without access to our audience. Getting that access means that we need the cooperation of all the parties, the FCC, the cable industry, and the consumer electronics manufacturers.

The FCC has the authority to resolve the regulatory issues, but the Congress must oversee this transition and take the necessary steps to make it happen.

I would like to close with some final thoughts. There was a dispute regarding the digital transmission standard. That dispute has been resolved. Also in the aggregate, we have more than 1,000 hours of high definition programming coming from our networks, with CBS providing virtually all of its prime time programming in HDTV. And we have encouraged all of the networks to do more.

Mr. Chairman, let us all work together to make sure the American public can watch the programming and enjoy the benefits of the wonderful digital service. Thank you, Mr. Chairman and members of the Committee.

[The prepared statement of Mr. Tucker follows:]

PREPARED STATEMENT OF BEN TUCKER, EXECUTIVE VICE PRESIDENT FOR BROADCAST OPERATIONS, FISHER BROADCASTING, INC., AND TELEVISION BOARD CHAIRMAN, NATIONAL ASSOCIATION OF BROADCASTERS

Thank you, Mr. Chairman, for the opportunity to appear before your Committee today to discuss the transition to digital television. My name is Ben Tucker. I am the Executive Vice President for Broadcast Operations for Fisher Broadcasting, Inc. I also am the National Association of Broadcasters (NAB) Television Board Chairman. I'm pleased to represent the broadcasting industry at this hearing.

Fisher Broadcasting, Inc. owns 12 television stations, the majority of which are licensed in the upper northwest states. We currently have two DTV stations on the air—KATU in Portland, OR and KOMO in Seattle, WA. DTV equipment is on order for the rest of our stations. I would like to highlight the fact that KOMO in Seattle currently provides local HDTV newscasts. As you can see, Fisher Broadcasting, Inc. is committed to making the DTV transition as quickly as possible. This commitment is the same for the entire broadcast industry.

Broadcasters Commitment to DTV

Stations on the Air

As of February 26, 2001, 182 DTV stations are on the air in 62 markets reaching 67.18 percent of all TV households across the nation.[1] Seventy-one of these stations—almost 40 percent—currently on the air are ahead of their required build-out schedule. These 182 DTV stations have met—or surpassed—the aggressive build-out schedule set by the FCC in order to meet the Congressional target date of 2006 to complete the digital transition.

Programming

The obvious advantage of DTV is the crisper pictures and enhanced viewing experience. Stations will be able to offer many more choices to consumers. Consumers will be the driving force behind the programming offered by DTV stations.

DTV stations are required to provide at least one free, over-the-air channel. This could come in the form of one high definition TV (HDTV) channel, or several streams of standard definition TV (SDTV) signals. Stations also could choose to offer some HDTV programming and some SDTV programming depending on the time of day and consumer demands. DTV stations are allowed to offer ancillary or supplemental services.[2]

The television networks currently offer hundreds of hours of HDTV programming. For example, CBS offers almost 1,000 hours per year, including nearly all prime time programming and major sporting events. ABC provides *NYPD Blue* and Disney films in HDTV. Locally, several stations—including Fisher Broadcasting's KOMO—provide local HDTV newscasts and a consortium of commercial stations exchange locally produced HDTV programs.

We are far ahead in the programming offerings in the DTV transition from those offered when the television industry transitioned to color. In the first year of color television back in the 1950s, only 68 hours were offered to viewers. With over 1,000 hours of HDTV programming this year, we are far outpacing the color TV rollout. That's good news because as the transition moves forward, we can only expect content providers will produce more and more programming in HDTV.

Even though there is consistent progress regarding programming and the number of DTV stations currently on the air, the transition still needs help with some major issues that threaten to throw the transition off the tracks.

---

[1] A list of stations currently on the air is attached as Exhibit A.
[2] Stations must pay a 5 percent fee on any profits earned from subscription services.

Broadcasters Call for Action

There are only 14 months left before the May 2002 deadline for all commercial stations to have a digital signal on the air. They face numerous obstacles from a regulatory standpoint, including the same build-out hurdles the existing 182 DTV stations faced.

What we have learned in the last few years is that we cannot accomplish this monumental task on our own. The transition to DTV is the biggest step for the television industry since the advent of color TV and represents a multi-million dollar expense for each individual station. Additionally, during the transition, each broadcast station will be operating essentially two stations, without any guarantee of additional revenue. Broadcasters are committed to this transition to bring DTV service to the American public. However, at this point, the DTV transition appears to be faltering due to several remaining issues that have yet to be resolved by all of the parties involved in this transition.

There are several entities that serve vital roles in this transition in addition to the broadcasting industry. In order for the transition to be successful, all parties must be willing do their part to get the job done.

The first party, the Federal Communications Commission (FCC), is charged with overseeing the implementation of DTV service to the American public. While the FCC has accomplished a great deal regarding the transition—including assigning an additional 1600 new DTV channel allotments—it has taken a hands-off approach with some of the remaining critical issues such as digital must-carry, DTV/cable interoperability, and DTV set standards. It is time for the FCC to take a leadership role in this transition and help focus all parties on getting the remaining pieces put in place so the goal of DTV can be realized as quickly as possible.

Cable operators, for example, have an important role in the transition. Nearly 70 percent of all homes receive over-the-air broadcast signals through cable providers. This means that cable operators hold an important key in the transition—access to viewers. A successful transition, after all, depends on consumers being able to see a broadcaster's digital product. Cable carriage of all over-the-air DTV channels and innovative digital services will create more demands for digital programming, resulting in consumers buying digital sets and converters at a faster pace, which helps drive the transition along.

Finally, consumers need the proper equipment to experience the benefits of DTV. This means that new DTV sets or set-top converters must first be manufactured and second, made available to the public. Consumers must be assured that the new digital products will work with cable set-top boxes and that the equipment can receive and decode DTV signals. Thus, manufacturers must work with cable companies to ensure that DTV sets are interoperable with digital cable boxes. Manufacturers must ensure that more DTV sets will include DTV tuners so consumers can receive the over-the-air signals.

The FCC has been relying on the marketplace to settle the remaining issues. We have learned that the marketplace is not driving the transition fast enough—placing the target date in jeopardy. We need resolution of the digital must-carry, DTV/cable interoperability, and DTV set reception issues or the transition will continue to falter and stall. I welcome the opportunity to outline these issues for you.

DTV Transmission Standard

Before discussing the other issues mentioned above, I would like to take the opportunity to dismiss any questions regarding the broadcasting industry's commitment to the FCC-approved DTV transmission standard, 8-VSB.

In the summer of 1999, concerns were raised among some in the broadcasting industry regarding the 8-VSB standard and its performance in urban markets and for mobile applications. Some believed that another transmission standard—COFDM—was more appropriate. When the issue was raised, most of the other entities involved in the transition accused the broadcasters of using it as a stalling tactic and questioned our commitment to DTV. We rose to this challenge and immediately took steps to resolve the issue.

In 2000, the broadcasting industry conducted a parallel investigation of VSB improvements and COFDM performance. This joint initiative included the National Association of Broadcasters (NAB) and Maximum Service Television (MSTV), with funding from the four networks (PBS in-kind), group broadcasters, and NAB.

Investigation of VSB included independent evaluations of second generation products and test performance in the field and improvements to the 8-VSB standard for possible modification of the standard to accommodate new applications. The project investigated the COFDM standard to test the performance of COFDM for existing and new services.

Upon completion of the testing in 2000, results were reported to the NAB and MSTV Boards of Directors in January 2001. After reviewing the results, both Boards passed a joint resolution that stated there is insufficient evidence to add COFDM as a DTV standard and thus it reaffirmed the commitment to the VSB standard.[3] Soon thereafter, the FCC affirmed the 8-VSB modulation system as the U.S. DTV transmission standard.

While virtually all of the broadcasting industry is now united behind the 8-VSB standard, DTV set reception must be improved. Broadcasters and, we hope, our manufacturer brethren are committed to seeing this happen post haste. Additionally, we are committed in helping to resolve the rest of the hurdles on this track to the DTV finish line.

DTV Must-Carry

Digital must-carry is the most important issue still facing the DTV transition. At this point, not many consumers can receive the currently available DTV signals via cable because cable, generally, will not talk to broadcasters about carriage of DTV signals. Must-carry of digital signals during the transition will help fuel the demand for digital programming, and will entice consumers to buy digital sets. Why should the 70 percent of Americans who are cable subscribers join the DTV transition by purchasing an expensive DTV set if they cannot easily get DTV broadcasts that are in their market?

The Communications Act of 1934, as amended by the Cable Act of 1992, mandates carriage of both analog and DTV signals.[4] The FCC is required to ensure the carriage of digital television signals;[5] however, it has so far failed to comply with this mandate. The FCC issued a *Notice of Proposed Rule Making* for digital must-carry in July 1998.[6] Nearly two-and-a-half years later, it issued a "partial" decision.[7] There, the FCC (1) refused to require dual must-carry of both analog and DTV signals; (2) asked for more information on channel capacity from cable operators; and (3) established that content to be carried after the transition is only one programming stream plus program related content.[8]

This partial decision does not solve the problems of the DTV transition—it only exacerbates them. Carriage of DTV signals during the transition is essential for a successful and timely conversion. Without must-carry, completing the transition even close to 2006 is impossible. The Congressional Budget Office recognized this in 1999 when it stated:

"The availability of digital programming on cable systems is a necessary, though not sufficient, condition for a timely transition. Without it, reaching the 85 percent penetration rate needed to end analog broadcasts in a market will take much longer because whenever the transition is completed, the largest number of households will probably be receiving DTV programming from cable providers." *Completing the Transition to Digital Television*, Congressional Budget Office Report, September 1999.

Even the FCC acknowledges cable carriage likely "is essential" to the DTV transition.[9] The question then remains—why does the FCC fail to take adequate steps to assure carriage on cable systems in order to facilitate the DTV transition?

Even after the transition is over, the FCC's decision on must-carry substantially cuts off consumers from realizing all the benefits of DTV. The FCC indicates it will require carriage of only one channel of each DTV broadcaster and other material "related" to that channel.[10] However, this completely dismisses the desirable choices broadcasters may offer to consumers by providing several SDTV signals (*i.e.*, multicasting). If a DTV station offers several free—but different—over-the-air programming choices, it should not be forced to choose which is the "main" program channel to be carried on the cable systems. Consumers should be offered all free broadcast programming through their cable system, regardless of whether that comes in the form of one HDTV channel or several SDTV channels, or a combination of both.[11]

---

[3] A copy of the Joint Resolution is attached as Exhibit B.

[4] Communications Act of 1934, § 614(a).

[5] *Id.* at § 614(b)(4)(B).

[6] *Notice of Proposed Rule Making*, CS Docket No. 98–120, July 10, 1998.

[7] *First Report and Order and Further Notice of Proposed Rule Making*, CS Docket No. 98–120, January 18, 2001 [hereinafter *First Report and Order*]

[8] *Id.* at ¶¶ 112 & 57.

[9] *See Fourth Further Notice of Proposed Rulemaking / Third Notice of Inquiry*, MM Docket No. 87-268, 10 FCC Rcd. 10540, 10542 (1995).

[10] *First Report and Order* at ¶ 112.

[11] Carriage of a "multicast" channel does not take up any more space on a cable system than a single HDTV channel. The same amount of space (19.4 megabits) is required. It makes no

The absence of digital must-carry frustrates Congressional intent in providing flexibility in the use of the spectrum to give consumers all the benefits of digital technology.

Finally, we have all heard the cries from the cable companies that digital must-carry will force them to take existing cable channels off their systems to make room for the DTV signals. These concerns are disingenuous. The broadcasting industry is not asking for an increase in the Cable Act's caps on the number of cable channels that must be devoted to broadcast channel carriage. Further, we do not ask for carriage of digital signals on smaller cable companies until they make their own transition to upgraded facilities and digital cable.

It is clear that cable companies are dramatically increasing their capacities, and will continue to do so with digital cable systems. In fact, at the height of the DTV transition when both analog and digital broadcast channels would be carried by cable systems,[12] the average analog cable system will have the capacity for approximately 130 channels.[13] An average digital cable system is predicted to have a capacity of 172 channels.[14] 1As a point of reference, the average capacity for cable systems in 1998 (when the FCC began its digital must-carry proceeding) was 75.[15]

As a final "nail in the coffin" on channel capacity concerns, at a FCC Cable Bureau hearing last year, the General Counsel of AT&T unwittingly but proudly professed that "[cable] channel capacity is not only increasing exponentially, but is about to go even beyond that as it [cable] goes digital."[16] He went on to say that AT&T's belief "is that we are going to be crying for content."[17] He had no answer when asked if that included digital must-carry signals.[18]

Digital must-carry is the most important, yet unresolved issue for the digital transition. The plain text of the must-carry statute is clear, cable operators "shall carry the signals" of broadcast operators.[19] We ask that Congress take every action necessary to ensure must-carry status for all digital broadcast channels during, as well as after, the transition.

DTV/Cable Interoperability

At this point, there are not standard DTV sets on the market that have connections that will work with digital cable set-top boxes.[20] Thus, there is no practical way for the 70 percent of consumers who view television via cable to get a broadcast DTV signal over cable today. Nor is there completion of the long promised built-to specs for cable ready DTV sets. Nor is there an indication that either will occur in time for the DTV transition to meet the Congressional deadlines.

There are incomplete, voluntary specifications between the consumer electronics and cable industries for DTV/Cable interoperability. Additionally, there is a remaining issue regarding copy protection for programming. All this translates into virtually no incentive for cable subscribers to purchase DTV receivers.

Agreements on these issues are both close and stalled. Quick resolution is needed to move the transition forward. This means there needs to be consumer-friendly IEEE 1394 connectors on all DTV receivers, set-top boxes and other DTV products and "cable-ready" characteristics for direct connection DTV receivers.[21]

For years, the broadcasting industry has been urging the FCC to mandate interoperability standards for DTV and cable products. At a minimum, it needs to secure strong manufacturer commitments for near-term provision of such products, or the

---

practical sense for cable companies not to allocate—at all times—enough space for a HDTV signal, which may follow or precede a multicast signal. It simply is not a space problem for cable to carry all free DTV channels sent from the broadcaster.

[12] In 2002, when all commercial broadcast stations must have a digital signal on the air, there would be an average of 12 broadcast channels carried. As the transition progresses, this number decreases back to the average of 6 broadcast channels at the end of the DTV transition. *See* NAB's Reply Comments in CS Docket No. 98–120, at Exhibit F (Dec. 22, 1998).

[13] *Id.*

[14] *Id.*

[15] *Id.*

[16] AT&T/Media One Cable Services Bureau Hearing, February 4, 2000.

[17] *Id.*

[18] Similarly, the Senior Vice President, Engineering and Technology for Media One cable has been quoted saying that "This digital capability effectively obliterat[es] the must-carry threat." Jim Barthold, Bandwidth Debate: Just How Much Will Be Enough (last modified Aug. 10, 1998). *http://www.mediacentrall.com/Magazines/CableWorld/News98/1998081003.html.*

[19] Communications Act of 1934, § 614(b)(1)(B).

[20] *See* DTV Products Chart, attached as Exhibit C.

[21] While copy protection issues must be soon settled, 1394 licensors should not be permitted to have a blanket ban on use of this copy protection technology for particular content, i.e. free broadcast programming.

transition will be further stalled. Again, Congress should take the necessary action to ensure resolution of these issues.

DTV Receiver Standards

The issue of receiver standards is important to the transition—this involves (1) mandating DTV tuners in all new TV sets sold, and (2) setting specific technical requirements regarding reception. Right now, if a consumer buys a DTV set, it is likely that the consumer will need to purchase an additional set-top box with a DTV tuner in order to receive DTV signals. Additionally, there is no guarantee that the DTV set will properly receive the over-the-air signals sent by broadcasters.

In the beginning of the DTV transition, the FCC set specific DTV transmission standards based on technical assumptions about receiver performance. The consumer electronics manufacturers have resisted any mandated receiver standards to meet the FCC's assumptions for reception. The FCC has relied on the marketplace to take care of this issue and has refused to set performance levels for DTV sets. It reaffirmed its position in January 2001. However, it turns out—as broadcasters had predicted—that early receiver performance does not match the FCC's assumptions. It is inconsistent for the FCC to expect to achieve certain DTV coverage and service goals, yet be unwilling to set performance levels for DTV sets. Why should consumers purchase DTV sets with poor reception performance?

By January 2001, there were approximately 780,000 DTV displays (with and without integrated tuners) sold to retailers. There are no breakout figures on sets with DTV tuners (integrated DTVs). At the same time, only 60,600 set-top tuner boxes were sold to retailers. Thus, there is only a small fraction of the hundreds of thousands of DTV displays that are able to receive a DTV signal over-the-air. At this rate, DTV receiver sales (integrated or set-top tuners) will not reach the penetration levels needed to complete the transition by the target date of 2006 set by Congress.

Broadcasters have urged the FCC to adopt All Channel Television Receiver Rules that will require that all new television receivers 13 inches and greater in screen size be capable of receiving all frequencies allocated by the FCC to television broadcasting, including all NTSC and all DTV channels.

While this is a significant step, it is not without precedent. The All Channel Receiver Act (47 U.S.C. § 303(s)) and the All Channel Television Receiver Rules,[22] provide the authority for such action by the FCC. These previous actions were taken to promote and develop the UHF frequencies. Congress, at that time, found that the lack of receivers capable of receiving UHF signals was the root of the problem for the faltering UHF service. It determined that "the only practical and effective means of insuring that such receivers get into the hands of the public is to enact legislation requiring that all sets manufactured are capable of receiving all of the channels allocated for television use."[23] This reasoning from the UHF situation applies to the current DTV situation—but now, with even more force.

In 1962, Congress determined that the dramatic step of the *All Channel Receiver Act* was necessary, even given initially increased costs (that would diminish with mass production). Congress reasoned that the small increase in cost was greatly offset by the benefits of "unlocking" the valuable UHF channels.[24] The same reasoning applies to the DTV transition today.

DTV is a unique transition of the entire television system to digital technology. Even though the price to consumers for an all-channel receiver will be higher than analog-only sets, the higher costs will be a small price to pay for "unlocking" the value of DTV channels for public benefit. Not to mention the fact that it also will release valuable NTSC channels, to be returned to the public for its benefit and use as Congress deems fit.

This bold action is necessary to revitalize a transition that has languished far too long. In January 2001, the FCC issued a *Further Notice of Proposed Rule Making* regarding this issue.[25] However, it only proposed to require tuners in sets that are 32 inches or larger, then phase-in tuner requirements for smaller sets. While this is a first step, it is not the bold action necessary to invigorate the DTV transition in order to meet Congress' 2006 timeframe. If necessary, Congress should take appropriate action to resolve these pending receiver issues.

---

[22] *First Report and Order*, All Channel Television Receiver Rules (All Channel Act), Docket No. 14760, 27 Fed. Reg. 11698 (Nov. 28, 1962).

[23] Senate Report No. 87–1526, 2d Sess. (1962), reprinted in 1962 U.S.C.C.A.N. Vol. 1, 1873.

[24] Id. at 1876.

[25] Report and Order and Further Notice of Proposed Rule Making, MM Docket No. 00-39, §§ 103-112 (January 18, 2001).

Other Build-Out Problems

As mentioned earlier, there are 14 months left before all commercial broadcasters must have a DTV signal on the air. There are approximately 1200 stations left to go on-air with DTV. Of the 182 DTV stations currently on the air, many faced build-out problems. These same problems, and more, will exist for the rest of the stations yet to make the transition.

Economic Issues

It costs approximately $8 million to $10 million to fully convert a station to digital operation. To date, the industry has spent hundreds of millions of dollars. Just to get a digital station on the air costs roughly $2 million. For many of the remaining stations and markets, these costs are well above the value of the existing analog station. And this, when there is no guarantee of any additional revenue from running two stations.

Tower Citing/Zoning Delays

New DTV stations require new DTV transmitting antennas. Stations must either use existing towers or build new towers. These changes often require approval from local zoning boards—which historically do not act quickly on these issues.

As part of the FCC's *Biennial Review* of the DTV transition, NAB conducted a survey of all commercial television stations asking specific questions about implementation problems. A surprising number of broadcasters (38.4 percent of respondents) reported that government—local and Federal—was causing delays in their digital rollout.[26] Stations cited numerous delays with local zoning or board approvals, the Federal Aviation Administration (FAA), local and Federal environmental agencies, as well as significant delays in the FCC approval process.

Once clearance is approved for any tower changes, the next hurdle for stations will be to find a tower crew to actually perform the work. There are limited numbers of tower companies with crews to do this specialized work. Further, as nearly 1200 stations place orders for the necessary DTV equipment, delivery delays from manufacturers are likely.

As you can see, merely getting a station on the air on schedule has its own difficulties, not to mention the larger regulatory issues that are threatening to hold up the DTV transition. Again, broadcasters are working toward the end, but there needs to be some help along the way from all parties involved, as previously discussed.

Conclusion

Mr. Chairman, it has been my great privilege to address this Committee on the subject of the digital television transition. I believe that broadcasters are fully committed to this transition that is poised to offer huge new benefits to the American public.

I hope that Congress will take a serious look at the issues facing the DTV transition and urge the cooperation of all parties to get the transition on a quicker pace toward completion.

---

[26] *See 2000 Digital Implementation Survey* , May 2000 (attached as an Exhibit to NAB's Comments in MM Docket No. 00–39, May 17, 2000).

EXHIBIT A—DTV STATIONS ON AIR BY STATE, 27-Feb-01

State	Calls	City Of License	Network	Group/Organization	DMA Name	DMA Rank
AL	WALA	Mobile	FOX	Emmis Communications Corp	Mobile-Pensacola	62
AZ	KNXV	Phoenix	ABC	Scripps Howard Broadcasting	Phoenix	17
	KPHO	Phoenix	CBS	Meredith Corp. Broadcasting	Phoenix	17
	KSAZ	Phoenix	FOX	Fox Television Stations Inc.	Phoenix	17
	KPNX	Phoenix	NBC	Gannett Broadcasting	Phoenix	17
	KUTP	Phoenix	United Paramount Net	Chris Craft/United Television	Phoenix	17
CA	KCOP	Hollywood	United Paramount Net	Chris Craft/United Television	Los Angeles	2
	KTLA	Hollywood	Warner Bros.	Turner Broadcasting Co.	Los Angeles	2
	KCAL	LA/Norwalk	Pure Independent	Young Broadcasting Inc.	Los Angeles	2
	KABC	Los Angeles	ABC	ABC Broadcasting	Los Angeles	2
	KCBS	Los Angeles	CBS	CBS Television Stations	Los Angeles	2
	KTTV	Los Angeles	FOX	Fox Television Stations Inc.	Los Angeles	2
	KNBC	Los Angeles	NBC	NBC Television Stations Div.	Los Angeles	2
	KCET	Los Angeles	PBS		Los Angeles	2
	KWHY	Los Angeles	Pure Independent		Los Angeles	2
	KTVU	Oakland	FOX	Cox Television	San Francisco-Oakland-San Jose	5
	KXTV	Sacramento	ABC	Gannett Broadcasting	Sacramento-Stockton-Modesto	19
	KTXL	Sacramento	FOX	Tribune Broadcasting Co.	Sacramento-Stockton-Modesto	19
	KCRA	Sacramento	NBC	Hearst-Argyle Television, Inc.	Sacramento-Stockton-Modesto	19
	KGTV	San Diego	ABC	McGraw-Hill Broadcasting Co.	San Diego	25
	KFMB	San Diego	CBS	Midwest Television Inc.	San Diego	25
	KNSD	San Diego	NBC	NBC Television Stations Div.	San Diego	25
	KSWB	San Diego	Warner Bros.	Tribune Broadcasting Co.	San Diego	25
	KGO	San Francisco	ABC	ABC Broadcast Group	San Francisco-Oakland-San Jose	5
	KPIX	San Francisco	CBS	CBS Television Stations	San Francisco-Oakland-San Jose	5
	KRON	San Francisco	NBC	Young Broadcasting Inc.	San Francisco-Oakland-San Jose	5
	KQED	San Francisco	PBS	KQED, Inc.	San Francisco-Oakland-San Jose	5
	KBHK	San Francisco	United Paramount Net	Chris Craft/United Television	San Francisco-Oakland-San Jose	5
	KBWB	San Francisco	Warner Bros.	Granite Broadcasting Corp.	San Francisco-Oakland-San Jose	5
	KICU	San Jose	Pure Independent	Cox Television	San Francisco-Oakland-San Jose	5
	KNTV	San Jose	Pure Independent	Granite Broadcasting Corp.	San Francisco-Oakland-San Jose	5
	KOVR	Stockton	CBS	Sinclair Broadcast Group Inc.	Sacramento-Stockton-Modesto	19
CO	KMGH	Denver	ABC	McGraw-Hill Broadcasting Co.	Denver	18
	KDVR	Denver	FOX	Fox Television Stations Inc.	Denver	18
	KRMA	Denver	PBS		Denver	18
CT	WFSB	Hartford	CBS	Meredith Corporation Broadcasting	Hartford & New Haven	27

	WTNH	New Haven	ABC	LIN Television Corporation	Hartford & New Haven	27
DC	WJLA	Washington	ABC	Allbritton Communications	Washington, DC	8
	WUSA	Washington	CBS	Gannett Broadcasting	Washington, DC	8
	WTTG	Washington	FOX	Fox Television Stations Inc.	Washington, DC	8
	WRC	Washington	NBC	NBC Television Stations Div.	Washington, DC	8
	WETA	Washington	PBS	Gtr. Wash. Educ. Telecomm.	Washington, DC	4
DE	WHYY	Wilmington	PBS	WHYY Incorporated	Philadelphia	52
FL	WTLV	Jacksonville	NBC	Gannett Broadcasting	Jacksonville, Brunswick	16
	WPLG	Miami	ABC	Post-Newsweek Stations, Inc.	Miami-Ft. Lauderdale	16
	WSVN	Miami	FOX	Sunbeam Television Corp.	Miami-Ft. Lauderdale	22
	WFTV	Orlando	ABC	Cox Television	Orlando-Daytona Beach-Melbourne	22
	WOFL	Orlando	FOX	Meredith Corporation Broad.	Orlando-Daytona Beach-Melbourne	13
	WTSP	St. Petersburg	CBS	Gannett Broadcasting	Tampa-St. Petersburg (Sarasota)	13
	WFTS	Tampa	ABC	Scripps Howard Broadcasting	Tampa-St. Petersburg (Sarasota)	13
	WTVT	Tampa	FOX	Fox Television Stations Inc.	Tampa-St. Petersburg (Sarasota)	13
	WFLA	Tampa	NBC	Media General Bcst. Group	Tampa-St. Petersburg (Sarasota)	10
GA	WSB	Atlanta	ABC	Cox Television	Atlanta	10
	WGCL	Atlanta	CBS	Meredith Corporation Broadcasting	Atlanta	10
	WAGA	Atlanta	FOX	Fox Television Stations Inc.	Atlanta	10
	WXIA	Atlanta	NBC	Gannett Broadcasting	Atlanta	115
	WRDW	Augusta	CBS	Gray Communications System	Augusta	71
HI	KHVO	Hilo	ABC	Hearst-Argyle Television, Inc.	Honolulu	71
	KITV	Honolulu	ABC	Hearst-Argyle Television, Inc.	Honolulu	71
	KMAU	Wailuku	ABC	Hearst-Argyle Television, Inc.	Honolulu	3
IL	WLS	Chicago	ABC	ABC Broadcast Group	Chicago	3
	WFLD	Chicago	FOX	Fox Television Stations Inc.	Chicago	3
	WMAQ	Chicago	NBC	NBC Television Stations Div.	Chicago	3
	WCPX	Chicago	Pax TV	Paxson Communications Corp.	Chicago	3
	WSNS	Chicago	Telemundo	Telemundo Group, Inc.	Chicago	3
	WGN	Chicago	Warner Bros.	Tribune Broadcasting Comp.	Chicago	161
	WGEM	Quincy	NBC	QNI Broadcast Group	Quincy-Hannibal-Keokuk	26
IN	WRTV	Indianapolis	ABC	McGraw-Hill Broadcasting Co.	Indianapolis	26
	WISH	Indianapolis	CBS	LIN Television Corporation	Indianapolis	26
	WXIN	Indianapolis	FOX	Tribune Broadcasting Co.	Indianapolis	26
	WTHR	Indianapolis	NBC	Dispatch Broadcast Group	Indianapolis	87
	WNDU	South Bend	NBC	Michiana Telecasting Corp.	South Bend-Elkhart	48
KY	WKPC	Louisville	PBS		Louisville	32
	WXIX	Newport	FOX	Raycom Media Inc.	Cincinnati	97
LA	WLPB	Baton Rouge	PBS		Baton Rouge	6
MA	WCVB	Boston	ABC	Hearst-Argyle Television, Inc	Boston	

EXHIBIT A—DTV STATIONS ON AIR BY STATE, 27-Feb-01—Continued

State	Calls	City Of License	Network	Group/Organization	DMA Name	DMA Rank
	WBZ	Boston	CBS	CBS Television Stations	Boston	6
	WFXT	Boston	FOX	Fox Television Stations Inc.	Boston	6
	WHDH	Boston	NBC	Sunbeam Television Corp.	Boston	6
	WHUB	Marlborough	Pure Independent	USA Broadcasting	Boston	6
	WGBY	Springfield	PBS	WGBH Educational Foundation	Springfield-Holyoke	105
MD	WMPT	Annapolis	PBS		Baltimore	24
	WMAR	Baltimore	ABC	Scripps Howard Broadcasting	Baltimore	24
	WJZ	Baltimore	CBS	CBS Television Stations	Baltimore	24
	WBFF	Baltimore	FOX	Sinclair Broadcast Group Inc.	Baltimore	24
	WBAL	Baltimore	NBC	Hearst-Argyle Television, Inc.	Baltimore	24
	WBOC	Salisbury	CBS		Salisbury	162
ME	WCBB	Augusta	PBS	Maine Public Broadcasting	Portland-Auburn	80
MI	WXYZ	Detroit	ABC	Scripps Howard Broadcasting	Detroit	9
	WWJ	Detroit	CBS	CBS Television Stations	Detroit	9
	WJBK	Detroit	FOX	Fox Television Stations Inc.	Detroit	9
	WDIV	Detroit	NBC	Post-Newsweek Stations, Inc.	Detroit	9
	WTVS	Detroit	PBS		Detroit	9
	WOOD	Grand Rapids	NBC	LIN Television Corporation	Grand Rapids-Kalamazoo-Battle Creek	38
	WKBD	Southfield	United Paramount Net	CBS Television Stations	Detroit	9
MN	KMSP	Minneapolis	United Paramount Net	Chris Craft/United Television	Minneapolis-St. Paul	14
	KTTC	Rochester	NBC	QNI Broadcast Group	Rochester-Mason City-Austin	153
	KSTP	St. Paul	ABC	Hubbard Television Group	Minneapolis-St. Paul	14
	KTCI	St. Paul	PBS		Minneapolis-St. Paul	14
MO	KCPT	Kansas City	PBS		Kansas City	31
	KDNL	St. Louis	ABC	Sinclair Broadcast Group Inc.	St. Louis	21
	KMOV	St. Louis	CBS	Belo Corporation	St. Louis	21
	KTVI	St. Louis	FOX	Fox Television Stations Inc.	St. Louis	21
	KSDK	St. Louis	NBC	Gannett Broadcasting	St. Louis	21
MS	WMPN	Jackson	PBS	Mississippi Authority for ET	Jackson, MS	89
NC	WSOC	Charlotte	ABC	Cox Television	Charlotte	28
	WBTV	Charlotte	CBS	Jefferson-Pilot Communication	Charlotte	28
	WCCB	Charlotte	FOX	Bahakel Communications, Lt.	Charlotte	28
	WCNC	Charlotte	NBC	Belo Corporation	Charlotte	28
	WTVD	Durham	ABC	ABC Broadcast Group	Raleigh-Durham	29
	WNCN	Goldsboro	NBC	NBC Television Stations Div.	Raleigh-Durham	29
	WCTI	New Bern	ABC	Lamco Communications, Inc.	Greenville-New Bern-Washington	106

	WRAL	Raleigh	CBS	Capitol Broadcasting Co., Inc	Raleigh-Durham	29
	WRAZ	Raleigh	FOX	Capitol Broadcasting Co., Inc.	Raleigh-Durham	29
NE	KMTV	Omaha	CBS	Emmis Communications Corp.	Omaha	73
NH	WENH	Durham	PBS		Boston	6
	WMUR	Manchester	ABC	Imes Stations	Boston	6
NJ	WNJT	Trenton	PBS	New Jersey Public Bcstg. Auth.	Philadelphia	4
NV	KLAS	Las Vegas	CBS	Landmark Broadcasting	Las Vegas	53
	KNPB	Reno	PBS		Reno	111
NY	WABC	New York	ABC	ABC Broadcast Group	New York	1
	WCBS	New York	CBS	CBS Television Stations	New York	1
	WNYW	New York	FOX	Fox Television Stations Inc.	New York	1
	WNBC	New York	NBC	NBC Television Stations Div.	New York	1
	WPIX	New York	Warner Bros.	Tribune Broadcasting Co.	New York	1
OH	WCPO	Cincinnati	ABC	Scripps Howard Broadcasting	Cincinnati	32
	WKRC	Cincinnati	CBS	Clear Channel Communications	Cincinnati	32
	WLWT	Cincinnati	NBC	Hearst-Argyle Television, Inc.	Cincinnati	32
	WEWS	Cleveland	ABC	Scripps Howard Broadcasting	Cleveland	15
	WJW	Cleveland	FOX	Fox Television Stations Inc.	Cleveland	15
	WKYC	Cleveland	NBC	Gannett Broadcasting	Cleveland	15
	WBNS	Columbus	CBS	Dispatch Broadcast Group	Columbus	34
	WMFD	Mansfield	Pure Independent	Meisse Broadcasting	Cleveland	15
	WOIO	Shaker Heights	CBS	Raycom Media Inc.	Cleveland	15
OK	KFOR	Oklahoma City	NBC	New York Times Co. Bcstg.	Oklahoma City	45
OR	KOAC	Corvallis	PBS	Oregon Public Broadcasting	Eugene	122
	KATU	Portland	ABC	Fisher Broadcasting Inc.	Portland	23
	KOIN	Portland	CBS	Emmis Communications Corp.	Portland	23
	KGW	Portland	NBC	Belo Corporation	Portland	23
	KOPB	Portland	PBS	Oregon Public Broadcasting	Portland	23
	KPTV	Portland	UPN	Chris Craft/United Television	Portland	23
PA	WLVT	Allentown	PBS		Philadelphia	4
	WFMZ	Allentown	Pure Independent	Maranatha Bcstg. Co. Inc.	Philadelphia	4
	WITF	Harrisburg	PBS		Harrisburg-Lancaster-Lebanon	46
	WPVI	Philadelphia	ABC	ABC Broadcast Group	Philadelphia	4
	KYW	Philadelphia	CBS	CBS Television Stations	Philadelphia	4
	WTXF	Philadelphia	FOX	Fox Television Stations Inc.	Philadelphia	4
	WCAU	Philadelphia	NBC	NBC Television Stations Div.	Philadelphia	4
	WTAE	Pittsburgh	ABC	Hearst-Argyle Television, Inc.	Pittsburgh	20
	KDKA	Pittsburgh	CBS	CBS Television Stations	Pittsburgh	20
	WPXI	Pittsburgh	NBC	Cox Television	Pittsburgh	20
	WVIA	Scranton/Wilkes Barre	PBS		Wilkes Barre-Scranton	51

EXHIBIT A—DTV STATIONS ON AIR BY STATE, 27-Feb-01—Continued

State	Calls	City Of License	Network	Group/Organization	DMA Name	DMA Rank
SC	WRLK	Columbia	PBS	South Carolina ETV Commission	Columbia	86
	WSPA	Spartanburg	CBS	Media General Bcst. Group	Greenville-Spartanburg-Asheville	35
TN	WKPT	Kingsport	ABC	Holston Valley Broadcasting	Tri-Cities, Tn-Va	92
TX	KXAN	Austin	NBC	LIN Television Corporation	Austin	61
	WFAA	Dallas	ABC	Belo Corporation	Dallas-Ft. Worth	7
	KDFW	Dallas	FOX	Fox Television Stations Inc.	Dallas-Ft. Worth	7
	KERA	Dallas	PBS	North TX Public Broadcasting	Dallas-Ft. Worth	7
	KDAF	Dallas	Warner Bros.	Tribune Broadcasting Co.	Dallas-Ft. Worth	7
	KTVT	Fort Worth	CBS	CBS Television Stations	Dallas-Ft. Worth	7
	KXAS	Fort Worth	NBC	NBC Television Stations Div.	Dallas-Ft. Worth	7
	KTXA	Fort Worth-Dallas	UPN	CBS Television Stations	Dallas-Ft. Worth	7
	KTRK	Houston	ABC	ABC Broadcast Group	Houston	11
	KHOU	Houston	CBS	Belo Corporation	Houston	11
	KRIV	Houston	FOX	Fox Television Stations Inc.	Houston	11
	KPRC	Houston	NBC	Post-Newsweek Stations, Inc.	Houston	11
UT	KBYU	Provo	PBS	Brigham Young University	Salt Lake City	36
	KTVX	Salt Lake City	ABC	Chris Craft/United Television	Salt Lake City	36
	KSL	Salt Lake City	NBC	Bonneville International Corp.	Salt Lake City	36
VA	WCYB	Bristol	NBC	Lamco Communications, Inc.	Tri-Cities, Tn-Va	92
WA	KOMO	Seattle	ABC	Fisher Broadcasting Inc.	Seattle-Tacoma	12
	KIRO	Seattle	CBS	Cox Television	Seattle-Tacoma	12
	KING	Seattle	NBC	Belo Corporation	Seattle-Tacoma	12
	KCTS	Seattle	PBS		Seattle-Tacoma	12
	KXLY	Spokane	ABC	Morgan Murphy Stations	Spokane	78
	KCPQ	Tacoma	FOX	Tribune Broadcasting Co.	Seattle-Tacoma	12
	KPDX	Vancouver	FOX	Meredith Corp. Broadcasting	Portland, OR	23
WI	WKOW	Madison	ABC	Shockley Communications Co.	Madison	85
	WISC	Madison	CBS	Morgan Murphy Stations	Madison	85
	WTMJ	Milwaukee	NBC	Journal Broadcast Group, Inc.	Milwaukee	33
	WMVS	Milwaukee	PBS		Milwaukee	33

EXHIBIT B

RESOLUTION OF THE MSTV BOARD OF DIRECTORS AND THE NAB TELEVISION BOARD

With the support of 30 major broadcast organizations and the oversight of technical committees consisting of some 25 engineers representing all major technical viewpoints, the broadcasting industry concluded a comprehensive, objective and expedited series of studies and tests to determine whether COMM should be added to the current 8-VSB standard.

We conclude that there is insufficient evidence to add COMM and we therefore reaffirm our endorsement of the VSB standard.

We also conclude that there is an urgent need for swift and dramatic improvement in the performance of the present U.S. digital television system.

We therefore will take all necessary steps to promote the rapid improvement of VSB technologies and other enhancements to digital television and direct the staffs to develop a plan and promptly submit it to the Boards.

Exhibit C

DTV Products by Greg Tarr

Integrated High-Definition TV Sets

(Digital Decoder included)

Brand	Model	Display Type	Screen Size	Aspect Ratio	Native Scan Formats	Scan Conversion	Line Doubling For NTSC?	Built-In AC-3 Decoder?	IEEE 1394 Digital Interface	Available	Suggested Retail Price
Daewoo	DSP-3060N	Pure Flat Direct View	30W"	16:9	1080i	All Formats>1080i	Yes	Yes	No	Now	$2,999.95
Hitachi	61HDX98B*	7" CRT Rear Projection	61W"	16:9	540p, 1080i	NTSC>540p 480i>540p 480p>540p 720p>1080i 1080i>1080i	Yes	Yes	No	Now	$5,999.99 (includes dish)
Hitachi	61HDX01W*	7" CRT Rear Projection	61W"	16:9	540p, 1080i	NTSC>540p 480i>540p 480p>540p 720p>1080i 1080i>1080i	Yes	Yes includes HD component inputs	No,	TBA	$5,999.99 (includes dish)
Konka	HD3098U	Direct View	30W"	16:9	1080i	All Formats>1080i	Yes	Yes	No	TBA	$3,499.99
Konka	HD3498U	Direct View	34W"	16:9	1080i	All Formats>1080i	Yes	Yes	No	TBA	TBA
Marantz	HD6400W	9" CRT Rear Projection	64W"	16:9	525p, 1080i	NTSC>525p 480i>1080i 480p>1080i 720p>1080i 1080i>1080i	Yes	Yes	No	Now	$11,999.99
Philips	64PH9905	9" CRT Rear Projection	64W"	16:9	525p, 1080i	NTSC>525p 480i>1080i 480p>1080i 720p>1080i 1080i>1080i	Yes	Yes	No	Now	$9,990.99
ProScan	PS65000*	7" CRT Rear Projection	65W"	16:9	540p, 1080i	NTSC/Vid/Component/480i>480p Component 480i>480p SD 480i/480p>540p ATSC HD 1080i/720p>1080i DIRECTV HD>1080i	Yes	Yes	No	Now	$5,299.99
ProScan	PS34000* Performax	Direct View	34W"	16:9	540p, 1080i	NTSC/Vid/Component/480i>480p Component 480i>480p SD 480i/480p>540p ATSC HD 1080i/720p>1080i DIRECTV HD>1080i	Yes	Yes	No	TBA	$3,499.99
ProScan	PS38000* Performax	Direct View	38W"	16:9	540p, 1080i	NTSC/Vid/Component/480i>480p Component 480i>480p SD 480i/480p>540p ATSC HD 1080i/720p>1080i DIRECTV HD>1080i	Yes	Yes	No	Now	$3,999.99
RCA	F38310* Performax	Direct View	38W"	16:9	540p, 1080i	NTSC/Vid/Component/480i>480p Component 480i>480p SD 480i/480p>540p ATSC HD 1080i/720p>1080i DIRECTV HD>1080i	Yes	Yes	No	Now	$3,799.99
RCA	P61310*	7" CRT Rear Projection	61W"	16:9	540p, 1080i	NTSC/Vid/Component/480i>480p Component 480i>480p SD 480i/480p>540p ATSC HD 1080i/720p>1080i DIRECTV HD>1080i	Yes	Yes	No	TBA	$4,999.99
Samsung (Tantus Digital)	HCJ555W	7" CRT Rear Projection	55W"	16:9	1080i	All Formats>1080i	Yes Includes HD Component	Yes	No	Now	$6,999.99
Samsung (Tantus Digital)	HCJ655W	9" CRT Rear Projection	65W"	16:9	1080i	All Formats>1080i	Yes Includes HD Component	Yes	No	Now	$9,999.95
Sony	KWP-65HD1	CRT Rear Projection	65W"	16:9	960i, 480p, 1080i	NTSC>960i 480i>960i 480p>480p 720p>1080i 1080i>1080i	Yes (DRC)	No, on-board ProLogic	No	Now	$9,999.99
Sony	KW-34HD1	Direct View	34W"	16:9	960i, 480p, 1080i	NTSC>960i 480i>960i 480p>480p 720p>1080i 1080i>1080i	Yes	Yes	No	Now	$6,499.99
Zenith Inteq	IQB64W10W	9" CRT Rear Projection	64W"	16:9	1080i	All Formats>1080i	Yes, Includes HD Component	Yes	No	Now	$9,999.99
Zenith Inteq	IQB56W10G	7" CRT Rear Projection	56W"	16:9	1080i	All Formats>1080i	Yes, Includes HD-compatible video,	Yes	No	Now	$8,499.99

*Also receives DirecTV standard and HD satellite programming

Courtesy of TWICE

DTV Products by Greg Tarr

Digital TV Set Top Decoders

Brand	Model	DTV Formats Received	Scan Conversion (Input>Output)	HD Interface for DTV Monitor	Includes NTSC Receiver	Includes NTSC Line Doubler/ Scaler	Built in Dolby Digital Decoder	IEEE 1394 Digital Interface	Available	Suggested Retail Price
EchoStar	Adapter module for Dish 5000**	Core 18 Table 3 Formats	Passes Undecoded HD Signal in Format Received	Connects via RF (RGB) To Separate DTV Decoder	No	No	No	No	Now	$299.95
EchoStar	6000**	Core 18 Table 3 Formats (with optional tuner module)	All Formats>480i All Formats>1080i All Formats>720p	HD Component, RGB H&V via BNC, RGB via 15-pin D-Sub	Has Upgradable Expansion Slot For Add-on Terrestrial NTSC/ATSC Tuner modules	Yes	Yes (2 Channel)	No	Now	$499.95; optional ATSC tuner module $100i
EchoStar	6100** (6000 with ATSC/NTSC tuner module)	Core 18 Table 3 Formats	All Formats>480i All Formats>1080i All Formats>720p	HD Component, RGB H&V via BNC, RGB via 15-pin D-Sub	Receives both NTSC and ATSC Terrestrial Broadcasts	Yes	Yes (2 Channel)	No	Oct-00	$599.95
General Instruments	HDD200*** Adapter module for 4DTV digital C-Band Decoder	C-Band; No ATSC tuner included	Selectable: All Formats>480p All Formats>720p All Formats >1080i	HD Component, RGBHV	No	Yes, multiplies externally tuned NTSC signals	No	No	Now	$1,599.99
Hughes Network Systems	HSYE-4686 Platinum HD*	Core 18 Table 3 Formats	All Formats>480i All Formats>1080i	HD Component	Yes	No	Yes	No	Oct-00	TBA
Integra	IT815ST Formerly Unity Motion HDR-1000A	Core 18 Table 3 Formats	HD Formats>1080i SD Formats>480p	HD Component, RGBHV D-sub	No	No	Yes	No	Now	$796.95
JVC	TU6000RU**	Core 18 Table 3 Formats (with optional tuner module)	All Formats>480i All Formats>1080i All Formats>720p	HD Component, RGB H&V via BNC, RGB via 15-pin D-Sub	Has Upgradable Expansion Slot For Add-on Terrestrial NTSC/ATSC Tuner modules	Yes	Yes (2 Channel)	No	Now	$499.95; optional ATSC tuner module $100
Konka	HD-0001	Core 18 Table 3 Formats	All Formats>1080i	RGB (High Density 15-pin D-sub)	No	No	Yes	No	TBA	$999.95
Loewe	HDT-100	Core 18 Table 3 Formats	Switchable: All Formats>480i All Formats>480p All Formats>720p All Formats>1080i	HD Component, RGB via VGA,	Yes	No	Yes (2-channel)	No	04-00	$1,599.99
Mitsubishi	SR-HD500*	Core 18 Table 3 Formats	All Formats>480i All Formats>1080i	HD Component Video RGB H&V-Sync	Yes	No	Yes (2-channel)	No	Oct-00	$1,025.00 includes dish
Mitsubishi	SR-HD400*	Core 18 Table 3 Formats	All Formats>480i All Formats>1080i	HD Component Video RGB H&V-Sync	Yes	No	Yes (2-channel)	No	Oct-00	$825.00 omits DirecTV dish
Panasonic	TU-HDS20*	Core 18 Table 3 Formats	All Formats> Any Output Selected	Switchable: HD Component or RGB H-V (RCA jacks)	Yes	Yes	No	No	Oct-00	$1,099.95
Panasonic	TU-DA2420	Dish for the TU-HDS20	N.A.	N.A.	N.A.	N.A.	N.A.	N.A.	Now	$199.95
Pioneer	SH-D505	Core 18 Table 3 Formats	Switchable: All Formats>480p All Formats>720p All Formats>1080i	HD Component Video, RGB H&V-Sync	No	No	No	Upgradable Expansion Port	Now	$2,499.99
Pioneer	SH-D09	Core 18 Table 3 Formats	Switchable: All Formats>480i, All Formats>1080i, All Formats>480p	Expansion Slot Connection for Pro-700HD	No	No	Yes	Upgradable Expansion Port	Now	$2,499.99
ProScan	PSHD105*	Core 18 Table 3 Formats	Switchable: All Formats>480i, All Formats>540p, 720p>1080i, 1080i>1080i	RGB via VGA	Yes	Yes	Yes	No	Now	$549.99

* Also receives DirecTV standard and HD satellite services
** Also receives Dish Network standard and HD satellite services
*** Receives HD signals from digital C-band satellites

Digital TV Set Top Decoders—continued from page 19

Brand	Model	DTV Formats Received	Scan Conversion (Input-Output)	HD Interface for DTV Monitor	Includes NTSC Receiver	Includes NTSC Line Doubler/ Scaler	Built in Dolby Digital Decoder	IEEE 1394 Digital Interface	Available	Suggested Retail Price
Proton	TBA	Core 18 Table 3 Formats	Switchable: All Formats>480i All Formats In Native Form, 720p>480p, 1080i>480p 480i>480p	RGB via VGA, Component (YUV)	No	No	Yes	No	TBA	TBA
RCA	DTC100*	Core 18 Table 3 Formats	Switchable: All Formats>480i, All Formats>540p, 720p>1080i, 1080i>1080i	RGB via VGA	Yes	Yes	Yes	No	Now	$549.99 without dish
Samsung	SIR-T100	Core 18 Table 3 Formats	Switchable: All Formats>480p All Formats>1080i	HD Component Video	No	Yes	Yes	No	Now	$1,999.99
Samsung	SIR-T150	Core 18 Table 3 Formats	Switchable: All Formats>480p All Formats>720p All Formats>1080i All Formats>NTSC Line doubles NTSC	HD Component RGB via VGA	No	No	No	No	Nov-00	$699.99
Samsung	SIR-TS200*	Core 18 Table 3 Formats	Switchable: All Formats>480p All Formats>720p All Formats>1080i All Formats>NTSC Line doubles NTSC	HD Component Video RBG via VGA	No	No	No	No	Fall-01	TBA
Sony	SAT-HD100*	Core 18 Table 3 Formats	Switchable: All Formats>480i All Formats>1080i	HD Component	Yes	No	No "DD compatible"	No	Q1-01	$849.99
Toshiba	DST-3000*	Core 18 Table 3 Formats	Switchable: All Formats>480i All Formats>1080i	HD Component Video	Yes	No	Yes	No	Oct-00	$999.99 w/o dish
Zenith	IQDTV-1080*	Core 18 Table 3 Formats	All Formats>1080i, 720p, 480p, 480i, All Formats>NTSC	HD Component RGB via VGA	Yes	Yes (enhanced)	Yes	No	Q1-01	TBA

* Also receives DirecTV standard and HD satellite services
** Also receives Dish Network standard and HD satellite services
*** Receives HD signals from digital C-band satellites

Courtesy of TWICE

The CHAIRMAN. Thank you very much.
Mr. Willner.

STATEMENT OF MICHAEL S. WILLNER, PRESIDENT AND CHIEF EXECUTIVE OFFICER, INSIGHT COMMUNICATIONS

Mr. WILLNER. Thank you, Mr. Chairman, Senators, good morning. I am Michael Willner. I am the President and CEO of Insight Communications, the eighth largest cable company in the United States with 1.4 million subscribers. I also serve as Vice-Chairman of the National Cable Television Association.

I would like to thank you for giving me the opportunity to tell a very exciting story about how digital television really is working in the marketplace. It is working not because of significant amounts of digital broadcasting by TV stations, but because of the cable industry voluntarily creating innovative and advanced new digital services.

Cable has moved with supersonic speed into the digital world, not because anybody told us to do so, but because our customers want us to. Today, 10 million households subscribe to digital cable. The key to cable's innovation has always been two-fold. Number one, consumer demand; Number two, freedom from excessive regulation—a combination absolutely critical in allowing us to raise the billions of dollars needed to invest in upgrading our technology. Our customers have been the ultimate winners because they now have more choices.

The cable industry's digital transition is happening with our own capital and without grants or subsidies from the government. Since the passage of the 1996 Telecommunications Act, cable has spent $42 billion to upgrade its infrastructure throughout the country.

Insight has been an industry leader in the development of new advanced digital services. By virtue of our investment we have added scores of new channels, developed interactive community news and information platforms, created video-on-demand services which electronically deliver up to 500 movie titles viewable whenever a viewer wants to watch them with full VCR functionality, made plans to open an electronic mall with up to 50 retailers, delivered lightning-fast access to the Internet, and we have recently launched our first facilities-based telephone service finally offering consumers a choice of local phone carriers.

To do all these things, our little company alone has invested nearly $500 million since the passage of the Act.

Broadcasters now argue that we should give to them large blocks of this newly created capacity to carry duplicative digital versions of their analog channels. This is to complete the digital transition that they committed to when Congress gave them about $70 billion worth of additional spectrum. Frankly, I thought the free grant of spectrum was valuable enough for them to have been motivated.

Cable is not seeking to hamstring a competitor here; we simply do not want to be the scapegoats for the broadcasters' problems.

The reality is that the cable industry is not just talking about the digital transition, we are doing something about it. Our rebuilds are 75 percent complete. Cable network HBO alone is offering more high definition programming than all of the broadcast networks combined. We have negotiated a technical standards agreement with the consumer electronics industry. We are developing a multitude of new digital channels and services. Our industry has committed to carry all of the digital equivalents of today's analog broadcast stations as soon as the broadcasters return their analog spectrum. And we have also committed to agree to carry the primary signal of broadcasters who return their analog spectrum earlier than the deadline and become digital-only broadcasters.

There should be no doubt about this. Cable wants to and will continue to provide customers complete access to the broadcast channels they enjoy today. But our customers do not want duplicative versions of each and every broadcast station. Dual must-carry is neither pro-consumer nor will it speed the digital transition because it does not encourage consumer migration to digital.

One scenario would have the broadcasters deliver digital signals to the cable operator, only to have them reconverted back into analog for delivery to a consumer's analog television set. The consumer would not see any difference and thus would not be encouraged to purchase a digital TV that exploits the medium's full potential.

The other scenario has cable systems retransmitting a digital version of the same programming consumers already receive in analog. The worst part about this scenario is that less than 1 percent of consumers would ever see it—only those few who have purchased very expensive digital TV sets.

The real problem here is that the broadcasters do not have a digital business plan developed. Most do not even know how much spectrum they will devote to free TV. Six years ago, this entire discussion was about the broadcasters' need to deliver high definition television over the air. Now, with the digital spectrum prize in hand, that plan seems dead.

Dual carriage would confiscate an additional 6 MHz of scarce channel capacity for programming services that do not even exist yet, and may never exist. Who benefits from that? Broadcasters. Why? Because in my view, they are seeking to block competition by occupying a second swath of bandwidth that otherwise could be used for new competitive rival services.

Consumers benefit if cable operators are free to use their digital capacity for the things that consumers want.

Mr. Chairman, there are a lot of different commercial interests here, broadcasters, cables, equipment manufacturers to name a few, but the most important is the public interest. The fact is that cable revenue comes directly from consumers. Therefore, cable operators must satisfy consumers' desires or risk losing them to our competitors. We respectfully submit that the public interest is best served by allowing cable the freedom to provide customers with new digital services that they want today and in the future, including those developed by broadcasters which are negotiated in a free and open marketplace. Thank you, sir.

[The prepared statement of Mr. Willner follows:]

PREPARED STATEMENT OF MICHAEL S. WILLNER, PRESIDENT AND CHIEF EXECUTIVE OFFICER, INSIGHT COMMUNICATIONS

Good morning. I am Michael Willner, President and CEO of Insight Communications. Insight serves approximately 1.4 million cable subscribers. I also serve as Vice-Chairman of NCTA. I would like to request that the NCTA paper on digital TV, which accompanies my testimony, be included in the record.

Thank you for giving me the opportunity to tell a very exciting story about how digital television REALLY is working in the marketplace. Working, not because of significant amounts of digital broadcasting but, because the cable industry has voluntarily created innovative, new advanced digital services.

Cable has moved with supersonic speed into the digital world not because anyone told us to but because our customers want us to. To date, 10 million digital cable boxes have been deployed. The key to innovation for cable has always been two-fold: consumer demand and freedom from excessive regulation—a combination absolutely critical in allowing us to raise the billions needed to complete our upgrades. Our customers have been the ultimate winners because they have more choices.

And it is important to note that the cable industry's digital transition is happening with our own capital, and without grants or subsidies from the government. Since the passage of the 1996 Telecommunications Act, cable has spent $42 billion dollars to upgrade its infrastructure.

Insight has been an industry leader in the development of new advanced services. By virtue of our investment, we:

- Added scores of channels; developed interactive community news and information services;
- Created video-on-demand services which electronically deliver up to 500 movie titles viewable whenever our customers want, with full VCR functionality;
- Plan to open an electronic mall with 50 retail outlets;
- Delivered high-speed Internet access; and launched our first facilities-based local telephone service offering consumers a choice of local phone carriers.

This is precisely what Congress intended in passing the Telecommunications Act 5 years ago.

To do all these things, our company alone has invested more than $500 million dollars since then.

In addition to carrying all of their analog signals, broadcasters now argue that we should give to them large blocks of this newly created capacity to carry duplicative digital versions of their analog channels. This is to complete the digital transition they committed to when Congress gave them $70 billion of additional spectrum. Frankly, I thought the free grant of that valuable spectrum was ample incentive.

Cable is not seeking to hamstring a competitor or to blame anyone. We simply don't want to be the scapegoats for broadcasters' problems.

The cable industry is not merely talking about the digital transition. Here's what we're doing to make it happen.

- Our digital rebuilds are 75 percent complete.
- HBO, Showtime and MSG are offering more high definition programming than all of the broadcast networks combined.
- We have negotiated a technical standards agreement with the consumer electronics industry so that new digital television sets will connect directly with cable systems.
- We are developing scores of new digital channels.
- Our industry has committed to carry the digital equivalent of today's analog broadcast stations when broadcasters return their analog spectrum.
- And to ensure consumers continued access to broadcast programming, we have agreed to carry the primary signal of broadcasters who return their analog spectrum and become digital-only broadcasters early.

Let there be absolutely no doubt about this—cable will continue to provide consumers complete access to the broadcast channels they enjoy today. But we are not prepared, nor do we believe the law requires us, to carry duplicative versions of each and every broadcast station.

Dual must carry is neither pro-consumer nor will it speed the digital transition because it does not encourage consumer migration to digital.

- One scenario would have the broadcasters deliver digital signals to the cable operator, only to have them converted into analog form for delivery to a consumer's analog TV set. The consumer would not see any difference and thus would not be encouraged to purchase digital TV sets that showcase the medium's promise.
- The other scenario has cable systems retransmitting the digital signal—which is largely a standard definition version of the same programming they receive in analog. This approach would take bandwidth away from digital services for millions of customers to deliver a duplicative broadcast channel to fewer than one half of 1 percent of consumers who have digital TV sets.

To date, broadcasters have not developed a digital business plan. They do not even know how much spectrum they will devote to "free TV." Six years ago this entire discussion was about broadcasters' need to deliver HDTV over the air. With that spectrum now in hand, that plan seems dead.

Dual carriage would appropriate an additional 6 MHz of scarce channel capacity for programming services that do not even exist yet, and may never exist. Even in newly rebuilt systems, cable bandwidth is far from unlimited. Reserving scarce capacity for one service inherently means that less spectrum is available for new interactive services or competing video services. Who benefits from that? Just broadcasters who would have reduced competition by occupying a second swath of bandwidth that could otherwise be used for new services.

Consumers benefit if cable operators are free to use their new digital capacity for the things consumers want. Whether they be broadcasters' digital services or other new services. In fact, several major broadcasters and cable MSOs already have signed agreements for carriage of both a broadcaster's analog and digital channels. But, these have been negotiated, not government mandated.

Mr. Chairman, there are many interests at stake here: broadcasters' interests, cable's interests, and equipment manufacturer's interests to name a few. But overarchingly, there is the public interest. Because cable's revenue comes directly from consumers, inherently, we must satisfy their desires or risk losing them to our competitors. We respectfully submit that the public interest is best served by allowing maximum flexibility for cable operators to provide consumers the new digital services they want today and in the future.

The Transition to Digital Television

PREPARED BY THE NATIONAL CABLE TELEVISION ASSOCIATION

Introduction

The transition to digitalis taking place in all sectors of the video distribution market. Digital television (DTV) technology has the capability to provide clearer and sharper, cinema-like pictures as well as CD-quality sound. It can also be used to compress video signals, allowing providers to offer multiple video programs in the same 6 MHz slot now occupied by one analog channel. Additionally, DTV technology can be used to provide new services such as data.

Broadcasters have expressed frustration about the ongoing transition to digital and they blame the Federal Communications Commission (FCC), the cable industry and the television set manufacturers for what they perceive as a lack of progress toward digital. They accuse the Commission of foot-dragging and criticize the cable and consumer electronics industries for not moving quickly enough to solve interoperability problems. They criticize cable operators and programmers for opposing a dual must-carry requirement—under which a cable operator would have to carry every broadcaster's analog and digital channels during the transition to digital.[1] And they complain that while broadcasters are honoring their end of the bargain, other parties to the process are not doing their part.

From listening to broadcasters, one would hardly guess that they asked for and received from the government a second 6 MHz channel of valuable spectrum free of charge to make this transition.[2] Contrary to their accusations, real progress is being made in the transition to digital.

To understand the issues and criticisms that have been raised, it is useful to briefly review the history of the digital transition and to outline the efforts of various industries to promote this transition.

Background: Broadcasters' Transition To Digital

In the late 1980s, high definition television (HDTV) was being advanced as the next great consumer-electronics breakthrough. The Japanese had developed an analog HDTV system that would offer consumers crystal-clear pictures and sound. It also required more than the 6 MHz of spectrum used by the existing analog TV system. Television set manufacturers saw HDTV as a way to sell more TV sets. Broadcasters saw it as a way to gain access to additional spectrum that otherwise might go to other users.[3]

---

[1] On January 23, 2001, the FCC released its *First Report and Order and Further Notice of Proposed Rule Making* on issues related to cable carriage of digital broadcast signals. In this decision, the FCC declined to impose a dual carriage requirement on cable operators. See p. 11 for further discussion of the FCC's decision.

[2] The FCC estimated that the total value of the digital spectrum ranged from $11 billion to $70 billion. Letters from Dr. Robert Pepper, Chief, Office of Plans and Policy, FCC, to Senators Lieberman, Kerrey, Conrad and Leahy, May 5, 1995.

[3] In the mid-1980s, manufacturers and users of two-way radios were pushing the FCC to allocate spectrum for land mobile uses. Police departments, ambulance services, commercial delivery companies and Motorola, which manufactured most of these radios, were trying to convince

Broadcasters petitioned the FCC to investigate the potential of advanced TV technology, while the U.S. Government urged the development of an American HDTV standard. Scientists and engineers from the public and private sectors began work on a new television system. In the meantime, the FCC began to examine the many issues involved in making a smooth transition to a new television system. Broadcasters started to lobby the government for a second channel of free spectrum.

While working on the HDTV standard, American electronics experts discovered that television programming could be digitized to transmit high-definition pictures. They also discovered that digital technology could be used to send multiple signals of "standard definition" (SDTV) programming in the same amount of spectrum. This digital standard—whether used to transmit HDTV or SDTV used just 6 MHz of spectrum instead of the 8 to 12 MHz used by the Japanese analog system. But, it was not compatible with the existing television system, meaning broadcasters would have to broadcast separate analog and digital signals during the transition to digital, and then return the analog spectrum at the end of the transition.

Broadcasters continued to urge the government to give them an additional 6 MHz of spectrum which, they argued, was necessary to make the transition to digital and to remain competitive. The government supported the broadcasters' arguments about the importance of making the transition. But not everyone agreed that they needed a second 6 MHz channel, or that they should get the additional spectrum for free. Opponents of this spectrum "giveaway" proposed giving the broadcasters only the amount of spectrum necessary to transmit a single standard definition digital signal and to make them pay for the additional spectrum.[4]

Broadcasters argued that if they didn't get the full 6 MHz, consumers would be deprived of one of the great benefits of digital technology—high definition television. In letters, speeches and testimony before congressional committees, broadcasters espoused the virtues of HDTV. The message was clear: they would use the digital spectrum to offer high definition television. An executive of the National Association of Broadcasters (NAB) said that TV stations "will use this spectrum for HDTV, pure and simple."[5]

In the end, broadcasters were granted 6 MHz of additional spectrum—valued at as much as $70 billion—free of charge. Despite their commitment to HDTV, broadcasters endorsed "spectrum flexibility" which would allow them to use the spectrum for other things. Broadcasters prevailed again and the government chose not to impose a HDTV requirement.[6] Congress put its stamp of approval on this plan in the 1996 Telecommunications Act.[7] Shortly thereafter, the FCC completed its 10-year-old proceeding on digital television by adopting a DTV table of allotments and establishing policies and rules for digital television.[8]

The FCC also established a digital television station buildout schedule and a target date of 2006 for cessation of analog broadcast service.[9] The FCC's rules required the TV stations affiliated with ABC, CBS, Fox and NBC in the top 10 markets to begin transmitting a digital signal by May 1, 1999.[10] By November 1, 1999, affiliates of these four broadcast networks in markets 11 to 30 were required to be on

---

the FCC that broadcasters were not using—and had no future use for—much of the spectrum allocated to them. Joel Brinkley, *Defining Vision*, Harcourt Brace & Company, 1997.

[4] *See e.g.*, Statement of Senator Bob Dole (R-KS), *Congressional Record*, April 17, 1996, p. S3443.

[5] Neil Hickey, "What's At Stake in the Spectrum War?" quoting NAB Executive Vice President Jim May, *Columbia Journalism Review*, July/August 1996.

[6] "To bolster DTV's chance for success, the Commission's decisions today allow broadcasters to use their channels according to their best business judgment, as long as they continue to offer free programming on which the public has come to rely." FCC Press Release on adoption of the *Fifth Report and Order*, MM Docket No. 87-268, April 3, 1997.

[7] 47 U.S.C. §336. Congress in 1996 limited eligibility for the new spectrum to incumbent broadcasters and permitted use of the spectrum for ancillary and supplementary services under certain circumstances.

[8] In the Matter of Advanced Television Systems and Their Impact Upon the Existing Television Broadcast Service, *Fifth Report and Order*, MM Docket No. 87-268, 12 FCC Rcd. 12809 (1997).

[9] Congress made the 2006 target date conditional in the Balanced Budget Act of 1997. Broadcasters need not give back their analog spectrum until 85 percent of the television households in their market are capable of receiving digital broadcasts, either over the air using a digital TV set or a digital to analog converter box, or through a multichannel video programming distributor.

[10] As of January 23, 2001, 35 of these 40 stations were on the air; the other stations have requested extensions to complete construction. *See* Top 10 Markets' DTV Status, at *www.fcc.gov*, February 6, 2001.

the air with a digital signal.[11] All other commercial stations must be transmitting a digital signal by May 2002, and non-commercial stations must do so by May 2003.[12]

Today, with more than 170 stations[13] transmitting a digital signal,[14] the broadcasters' early commitment to HDTV seems long forgotten.[15] Most broadcasters have made no major commitments to use the digital channels primarily to provide HDTV.[16] While some broadcasters are providing a limited amount of HDTV programming, many other broadcasters seem poised to provide non-HDTV, standard definition video programming over some of the additional spectrum, and to use the rest of their spectrum to provide commercial applications. Several broadcasters, for example, are planning to pool some of their digital spectrum to distribute data. Station groups are joining together to pursue this opportunity. iBlast, a consortium of large station groups, plans to create an infrastructure for delivering multimedia content to stations.[17] Another consortium of station groups, Broadcasters' Digital Cooperative, plans to lease part of their digital spectrum to support wireless data services.[18]

Even with these developments, many broadcasters are still unsure about what to do with their digital spectrum. Broadcast network officials have expressed uncertainty about their digital plans and acknowledge that they still have many questions about how to proceed. This uncertainty about the ultimate uses of the digital broadcast spectrum has not gone unnoticed at the FCC. At the NAB's annual convention in April 2000, former FCC Chairman William Kennard said he was reluctant to impose a dual carriage requirement on cable operators—that is, require op-

---

[11] Fifty-seven of these 80 stations were on the air as of January 23, 2001; the others have requested extensions until May 1, 2001 to complete construction. *See* Top 11-30 Markets' DTV Status, at *www.fcc.gov*, February 6, 2001.

[12] A NAB poll shows that 70 percent of station owners favor a delay in this rollout schedule. *Broadcasting & Cable*, April 10, 2000, p. 31. While many broadcasters are reluctant to publicly ask for a delay, Paxson Commumcations Chairman Lowell "Bud" Paxson has made it clear that he doesn't think the 2002 deadline for going digital is realistic: "We ain't going to make May of '02," "Delaying Digital TV," *Broadcasting & Cable*, January 29, 2001, p. 59.

[13] "177 TV Stations Broadcasting in Digital," NAB News Release, January 22, 2001.

[14] It is unclear exactly what broadcast stations are transmitting using these digital signals. In some cases digital TV transmitters may only be turned on for certain hours of the day. The programming on these channels, beyond a limited number of high definition programs, often consists of tape loops of high definition promos and upconverted standard definition analog signals. *See e.g.*, websites of KRON (San Francisco) at *www.kron.com* and KCTS (Seattle) at *www.kcts.org*.

[15] Some Members of Congress have expressed concerns about the broadcasters' move away from HDTV. In a September 1997 hearing before the Senate Commerce Committee, Senator Conrad Burns (R-MT) expressed concern about broadcasters' public statements about HDTV: "It would be unfortunate indeed if a stunning advance in technology made possible by American expertise—digital HDTV—would fail to be made widely available to our own citizens." House Telecommunications Subcommittee Chairman Billy Tauzin (R-LA) said that broadcasters' failure to provide HDTV "would violate the spirit of that unwritten agreement." "Datacasting Plans Raise Govt. Concerns About HDTV," *Communications Daily*, April 6, 2000, p. 4. At a July 2000 House Telecommunications Subcommittee hearing, Tauzin said that without widespread HDTV, Congress would reconsider its decision to give every TV station a digital channel at no charge. "Congress Disses Datacasting," *Broadcasting & Cable*, July 31, 2000, p. 17.

[16] CBS has taken the lead among the broadcast networks in providing high definition programming. It is originating most of its primetime entertainment programming and major sporting events in HD. ABC began broadcasting "NYPD Blue" in HD in January 2001 and broadcasts some HDTV movies on Saturday and Sunday nights. NBC airs 'The Tonight Show" in HDTV. Television set manufacturers are, in some cases, underwriting the production of high definition programming and these sponsorship agreements influence what programming is provided in HD.

[17] "Getting Together Over Data," *Broadcasting & Cable*, March 27, 2000, p. 6. There are 225 iBlast television stations including Tribune Broadcasting, Gannett Broadcasting, Cox Broadcasting, Post-Newsweek Stations, E.W. Scripps, Meredith Broadcasting, Media General Broadcasting, Lee Enterprises, The New York Times Co., McGraw-Hill Broadcasting, Smith Broadcasting, Northwest Broadcasting, Bahakel Communications, Bonneville International, Cosmos Broadcasting, Emmis Communications, Evening Post Publishing, Gray Communications, and Raycom Media. See *www.iblast.com,* February 6, 2001.

[18] *Id.* Station groups participating in the Broadcasters' Digital Cooperative include Granite Broadcasting, Benedek Broadcasting, Capitol Broadcasting, Citadel Communications, Clear Channel Television, Cosmos Broadcasting, Morgan Murphy Stations, Gray Communications, Nexstar Broadcasting, Pappas Telecasting, Paxson Communications, and Sunbelt Communications.

In an interview with *Broadcasting & Cable* magazine, Commerce Committee Chairman John McCain (R-AZ) said that if broadcasters leased their digital spectrum for data services rather than using it all themselves "it would be in direct contradiction to the commitment that they (the broadcasters) made when they got the spectrum." "Straight Telecom Talk," *Broadcasting & Cable*, July 24, 2000, p. 24.

erators to carry both a broadcaster's digital and analog signal—with broadcasters' digital business plans in a State of flux.[19]

In addition to the questions about how broadcasters will use the digital spectrum, there have been nagging concerns about the digital transmission standard. Numerous broadcasters, led by Sinclair Broadcasting Group, urged a review of the standard adopted by the FCC as the transmission standard for digital broadcast television signals, called 8-VSB.[20] Sinclair asked the FCC to allow broadcasters to transmit their digital signals using an alternative modulation method, called COFDM, in addition to 8-VSB.[21] Sinclair argued that COFDM overcomes several shortcomings of the 8-VSB technology, including difficulty with indoor reception.

In February 2000, the FCC rejected the Sinclair petition.[22] The Commission *reaffirmed* the 8-VSB modulation system in a January 2001 review of the digital television transition.[23] In this decision, the FCC said "based on our review of the record, the demonstrated improvements in DTV receiver performance, and the findings and recommendations of the industry, we find that there is no reason to revisit our decision to deny Sinclair's petition."[24]

The Consumer Electronics Industry applauded the Commission's decision saying its action, coupled with the recommendation of the broadcast industry," ends the debate over the DTV modulation standard."[25] Since completion of the industry tests, however, questions have arisen about the testing methodology.[26] Time will tell whether these recent actions will finally bring closure to the modulation debate, and whether greater certainty in this area is sufficient to spur additional commitments by television set manufacturers to roll out the next generation of DTV sets.[27] Broadcasters have embarked down the road to digital and, not surprisingly, some obstacles have arisen. But they began this journey at their own request and with a government grant of a second free channel of valuable spectrum.[28] Given the issues outlined above, it is disingenuous for broadcasters to try to shift the blame for the pace of the digital transition to others. The fact is that these other industries are making significant strides to promote the transition.

Digital Rollout by The Cable Industry

The cable industry has been a leader in the transition to digital and has taken on this role without government mandate or subsidy. Cable operators and programmers are working in a number of areas to ensure cable customers have access to new and unique digital services. Progress is being made in each of these critical areas:

- Cable plant upgrades that allow operators to offer new digital services;
- Creation of unique digital and high definition cable programming;
- Negotiation of retransmission consent agreements to make digital broadcast programming available to cable customers;
- Agreement between the cable industry and the consumer electronics industry to ensure digital TV sets work with cable systems.

Each of these areas is discussed in more detail below.

Cable Industry Upgrades

Cable operators have invested more than $42 billion since 1996 to upgrade their facilities in order to offer consumers new services, including digital cable. Digital

---

[19] "Kennard Seeking Channel-Space Data," *Multichannel News Online*, April 11, 2000.

[20] Other broadcasters joined in calling for a reexamination of the standard. For example, in their DTV biennial review comments, ABC and NBC said "[o]ur real world experience in receiving the 8VSB signal from our DTV stations is that this method of transmission does not provide reliable reception to our viewers." Joint Comments of NBC and The Walt Disney Company, MM Docket No. 00-39, filed June 16, 2000.

[21] More than 400 commercial and public stations supported the Sinclair petition.

[22] Letter from Magalie Roman Salas, Secretary, FCC, to Martin R. Leader, Fisher, Wayland, Cooper, Leader & Zaragoza, Counsel to Sinclair Broadcast Group, Inc., February 4, 2000.

[23] *Report and Order and Further Notice of Proposed Rulemaking*, MM Docket No. 00-39, January 19, 2001, 92.

[24] *Id.*, § 92.

[25] "CEA Applauds FCC Actions on DTV Transition," CEA News Release, January 19, 2001.

[26] "DTV Data is Called Invalid," *Broadcasting & Cable*, February 5, 2001, p. 38.

[27] Past uncertainty about the modulation technology did cause some manufacturers to delay rollout of the new DTV sets. For example, in June 2000, Sony announced just such a delay, acknowledging that review of the transmission method was one factor in its decision. "Sony Says Its Line of Digital HDTVs Will be Delayed," *Wall Street Journal*, June 8, 2000.

[28] *See e.g.*, "FCC's Chief Blasts Broadcasters for Delays in Digital-TV Shift," *Wall Street Journal*, October 11, 2000. "Basically, the broadcast networks were the beneficiaries of the biggest government giveaway since Peter Stuyvesant bought Manhattan from the Indians for $24," Mr. Kennard said, adding that "the networks' business model for the next decade can be summed up by the slogan of Twix candy bars: 'Two for me, none for you!'"

video service provides increased channel capacity through compression of four to 12 digital video signals in the same 6 MHz slot previously occupied by a single analog channel. As a result, cable customers are able to receive dozens of new programming services.

Consumers are responding by signing up for digital tiers in record numbers. To date, cable systems have attracted about 10 million digital customers. A survey released in March 2000 by the Cable and Telecommunications Association for Marketing (CTAM) showed impressive positive customer response to their upgraded, digital cable offerings. Of nearly 2,600 consumers polled, 95 percent expressed satisfaction with their service.[29]

Cable's New Digital and High Definition Programming

Program networks have already launched some 60 new digital channels offering consumers additional choice and further program diversity. Examples include the Biography Channel and History Channel International (from A&E); Science, Civilization and Kids (from Discovery); Noggin, Nick Too and Nickelodeon Games & Sports (from Nickelodeon); and style. (from E!). There are six new Hispanic channels from Liberty Cañales, new music channels from MTV and BET, and separate channels targeting Indian, Italian, Arabic, Filipino, French, South Asian and Chinese viewers from The International Channel. There are also many new premium offerings from HBO (HBO Family, ActionMAX and ThrillerMAX), Showtime (Showtime Extreme, Showtime Beyond) and Starz! Encore (Starz! Family, Cinema, Movies for the Soul, Adventure Zone).

Moreover, cable programmers are ahead of broadcasters on the high definition front. HBO is providing more HDTV programming in any given week than all of the broadcast networks combined. Showtime, Madison Square Garden, A&E and Discovery are also producing high definition programming. This is just the kind of high quality programming that will facilitate the transition to digital by enticing people to buy DTV sets.

Digital Must-Carry & Retransmission Consent

Broadcasters argue that a critical component of the digital transition is a government mandate that cable operators carry their digital signal as well as their analog signal during the transition to an all-digital broadcast system.[30] They contend that such a requirement is the only way broadcasters' digital programming will be carried on cable systems.[31] The cable industry strongly opposes—on legal, constitutional and policy grounds—a government-imposed requirement that cable operators carry both the analog and digital signals of every broadcaster.[32] Government-mandated dual must-carry unduly burdens cable operators and programmers, does not enhance but decreases program diversity and, therefore, does not serve the public interest.[33] The FCC recently expressed its own reservations about a dual must-carry requirement and tentatively concluded that such a requirement would be unconstitutional.[34] In its decision, the Commission said:

"Based on the existing record evidence, a dual carriage requirement appears to burden cable operators' First Amendment interests substantially more than is necessary to further the government's substantial interests of preserving the benefits of free over-the-air local broadcast television; promoting the widespread dissemina-

---

[29] CTAM's 1999 Digital Cable TV Customer Satisfaction Study.

[30] Broadcasters analog signals are carried on cable systems pursuant to the provisions of the 1992 Cable Act. Broadcasters can elect either mandatory cable carriage ("must-carry") or seek compensation for carriage of their signal ("retransmission consent") from cable operators in their coverage area.

[31] In fact, broadcasters want operators to carry more than just their digital television signals. National Association of Broadcasters President Eddie Fritts told attendees at the NAB's annual convention that operators should also have to carry broadcasters' Internet and other data services: "That means carrying the entire bitstream, not eliminat[ing] our new competitive data systems." "Fritts Says Cable Should Carry all of Local Broadcasters," *Cableday*, April 11, 2000.

[32] *See* Comments of the National Cable Television Association, CS Docket No. 98-120, filed October 13, 1998, and Reply Comments filed December 22, 1998.

[33] Giving two signals of every broadcast station preferential carriage over all cable networks is particularly disturbing when cable companies are the ones playing an increasingly prominent role in providing local programming and serving the interests of children—areas increasingly ignored by many broadcasters. *See e.g.* comments of senior VP of operations at Emmis Communications: "We think the government should have no say in what we do for children. It's a terrible financial business for us, and we don't think the government should tell us to run 3 hours of kids programming." "Who Decides What's Good for Children?," *Broadcasting & Cable*, January 29, 2001, p. 35.

[34] *First Report and Order and Further Notice of Proposed Rule Making*, CS Docket No. 98-120, January 23, 2001 at § 3.

tion of information from a multiplicity of sources; and promoting fair competition in the market for television programming."[35]

The Commission also said that after a broadcaster returns its analog channel and transmits only in digital format, the digital station will have must-carry rights—but only with respect to the "primary video" of the digital signal. According to the FCC, "primary video" means a single programming stream, along with any material related to that programming. New stations broadcasting only a digital signal are also entitled to carriage of the primary video of that signal.

The FCC also addressed technical and legal questions regarding the manner in which digital stations are to be carried by cable systems, pursuant to must-carry obligations or retransmission consent. Finally, the FCC adopted a Further Notice of Proposed Rulemaking in which it asked for comment on sonic of its tentative conclusions and related digital must-carry issues.[36]

The FCC was right to decide not to impose a dual carriage requirement on cable operators. The facts belie the broadcasters' claim that such a government mandate is necessary. The marketplace is working to resolve digital carriage issues. The cable industry will carry broadcasters' primary digital signal at the end of the transition, and will continue to carry their analog signals during the transition. No broadcaster will lose its voice, nor will any consumer lose access to his or her favorite broadcast channel.

In addition, large cable MSOs have entered into retransmission consent agreements with some broadcast station owners to carry digital broadcast programming during the transition. For example, AOL Time Warner has entered into comprehensive agreements for carriage of the digital signals of the four major broadcast networks, several station group owners, and a group of public broadcasters. AT&T has digital carriage agreements with Fox and NBC, and continues discussions with other broadcasters. Other negotiations are underway between broadcast and cable companies and are likely to yield additional agreements for the carriage of broadcast stations' digital signals on cable systems.

Like all aspects of the digital transition, these discussions take time.[37] But, the marketplace is working to resolve the digital carriage issue. A government-imposed digital must-carry rule will in no way provide consumers with an incentive to buy new digital television sets. Instead of arguing for such a requirement, broadcasters can provide this incentive by developing distinct and compelling programming that consumers want to watch. The retransmission consent agreements that have been reached and the ongoing discussions between cable and broadcast companies validate that as cable companies add channel capacity, and as broadcasters develop specific digital programming that consumers want, cable companies will carry such programming.[38]

Compatibility Issues

Another area where progress has been made to ensure a smooth transition to digital is the compatibility between cable systems, set-top boxes and digital television (DTV) sets. The cable industry has addressed the issue of compatibility and solutions are available. Cable systems deliver high definition digital signals to DTV sets by using so-called "component analog" connectors between a cable set-top box and a DTV set. In some cases, content providers may require copy protection before they will make high quality digital programming available to cable.

There are two approaches by which DTV sets can be connected to cable with adequate copy protection and security. First, an HDTV-capable set-top box can be connected to a DTV using a digital interface or connection, such as a "1394/5C"[39] or functionally equivalent digital link. This digital link will include copy protection technology to ensure that the digital signals cannot be pirated as they cross between the set-top box and the DTV set. Second, the functionality of the set-top box can he incorporated within the digital television itself. Using this approach, the DTV set connects directly to the cable system without the need of a set-top box. Since there

---

[35] *Id.*

[36] In conjunction with the *Further Notice*, the FCC sent out a survey to 16 cable operators seeking information on channel capacity and retransmission consent agreements to carry digital signals.

[37] Retransmission consent discussions have in some cases been hampered by the fact that many broadcasters do not have in place definite business plans for their digital spectrum.

[38] See e.g. "Time Warner Adds HDTV in Houston," *Multichannel News*, November 27, 2000, p. 22: "High-Definition TV: So Close, and Yet So Far," *Newsday*, January 10, 2001, p. C5.

[39] Several companies have developed the "SC" Digital Transmission Content Protection (DTCP) technology. Use of DTCP has been subject to ongoing discussion and the negotiation of terms and conditions between equipment manufacturers and content providers. The cable industry supports the proposed SC technology as an effective way to provide copy protection.

is no set-top box and, therefore, no extended connection to the DTV set, there is no opportunity for the digital signal to be stolen and copied.

Both of these approaches required inter-industry technical discussions and consensus. The cable and consumer electronics industries worked diligently to resolve these outstanding technical issues. In December 1998, the cable industry and the Consumer Electronics Association (CEA) agreed to the necessary changes in the IEEE 1394 specification to promote compatibility between digital television receivers and digital set-top boxes.[40]

Beginning in July 1999, the cable and consumer electronics industries conducted a series of joint meetings to address additional compatibility issues between cable systems and consumer electronics equipment. As a result of these meetings, three significant agreements were reached.

On February 23, 2000, CEA and NCTA announced two voluntary agreements to allow future consumer digital television sets and digital cable systems to work together. The agreements detail the technical specifications that will enable consumers to receive DTV programming and services over cable systems.[41] The first agreement details the technical specifications that will allow DTV receivers to connect to cable television systems. This agreement assures a cable customer who buys a DTV set that the set can be connected directly to his or her cable wire. The second agreement spells out how systems will transmit Program and System Information Protocol ("PSIP") data—the raw material provided by broadcasters and cable programmers that is used to make up electronic program guides created in a TV set.

These two technical agreements allow manufacturers to proceed with the production of digital TV receivers built to the agreed-upon technical specifications.[42]

On May 24, 2000, NCTA and CEA announced a third agreement to aid consumers in their purchase of new digital television equipment. This agreement established the labeling to be used to inform consumers about various digital television sets' capabilities to receive digital and interactive digital TV services. However, on September 15, 2000, the FCC, acting on a number of issues regarding cable and the digital television transition, instead required a different set of labels—using the term "cable ready"[43]—for digital television sets to indicate their capability to operate with cable television systems.

The CHAIRMAN. Thank you very much, Mr. Willner.

Mr. Sagansky, what do you expect Paxson Communications to look like in 10 years? Do you still expect to be a broadcaster?

Mr. SAGANSKY. Yes, we do expect to be a broadcaster, and we expect to be a broadcaster which will be multicasting multiple channels of high quality DTV family programming. And you know, as we look out to the future, we think this is the best possible use of our spectrum. It is not datacasting and it is not HDTV, because we have said all along that HDTV is fantastic for sporting events and movies, and other kinds of programming that take up a lot of bandwidth. But for us, for our family programming, even a standard DTV signal is a huge improvement over what consumers are seeing today.

---

[40] Through the efforts of CableLabs and its OpenCable project, the cable industry developed specifications for cable set-top boxes that could be sold at retail stores. The security features of these boxes would be included in a separate security module—a "Point-of-Deployment" or "POD" module—to be obtained from the cable operator. On December 15, 2000, CableLabs submitted a final "PHI" (POD-HOST Interface) license agreement to the FCC. The technology provided by this license is used to ensure security and to facilitate copy protection of high quality digital content as it passes across the interface between the POD module and the cable set-top box. This license will allow equipment manufacturers to begin producing digital set top boxes to be sold at retail.

[41] *See* Letter from Robert Sachs, President and CEO, NCTA, and Gary Shapiro, President and CEO, CEA, to Chairman Bill Kennard, FCC, filed February 22, 2000.

[42] DTV sets built to these specifications are likely to be available in retail stores within 14 to 18 months.

[43] *Report and Order*, PP Docket No. 00-67, September 15, 2000. The Commission ordered that digital television sets that work with cable but that do not have a 1394 connector—therefore limiting the sets to one-way capability—will be labeled "Digital Cable Ready 1." DTV sets with a 1394 connector will be labeled "Digital Cable Ready 2." DTV sets that have set-top functionality integrated into the sets—and therefore do not need a 1394 connector to work with two-way cable services—will be labeled "Digital Cable Ready 3."

The CHAIRMAN. It has been estimated that it costs between $2 and $8 million dollars to convert a broadcast TV station to digital. How much would you estimate that your company has spent on converting analog TV stations to DTV or HDTV?

Mr. SAGANSKY. Well, so far this year alone, we are going to spend $25 million, and it is going to cost us $100 million by the time that we are done.

The CHAIRMAN. How much have you spent?

Mr. SAGANSKY. We have spent $25 million.

The CHAIRMAN. Paxson Communications has been a vocal advocate of broadcasters being paid to vacate the spectrum early, effectively trying to sell the spectrum to other telecommunications concerns at a premium in return for vacating the spectrum early. Does this not presume that the broadcasters own the spectrum, while in reality it belongs to the American public, who has been loaning it to them based on the promise of digital television?

Mr. SAGANSKY. Senator, I think you are talking about channels, the 700 MHz options which are coming up in September, and those are channels 60 to 69. We actually have 18 stations that are operating in that spectrum. And we know that once we have to vacate for the use of wireless advanced technology, we are going to lose our entire over-the-air audience, because there is no digital television reception out there.

So what we are trying to do is we are working with a group of incumbent broadcasters, who are all licensed on that 59 to 69 band, and we are formulating a plan in accordance with the FCC's direction to clear this 700 MHz band prior to the end of the digital transition period. The plan which will be finalized, we are working on it now, is going to be presented to all incumbent broadcasters as well as potential bidders in the FCC auction scheduled for this coming September.

And the basis of the plan that we are developing has shown a lot of interest from the FCC. Discussions are ongoing with the FCC and among incumbent broadcasters but you know, we are hopeful that our initiative will clear the band and lead to a successful auction.

The CHAIRMAN. Mr. Tucker, when Congress gave you this free spectrum, you pulled a classic bait and switch. In October of last year, William Safire in the *New York Times* wrote: "The broadcasters insisted that the airwaves were their entitlement. With the gift of the new spectrum, they promised to deliver free TV broadcasts on high definition television." Yet in January of this year, former NAB executive vice president John Abel said, quote: "Broadcasters shouldn't have to worry about HDTV, there's nothing in the 1996 Telecommunication Act requiring HDTV or forbidding data."

What I would like to know is what were your original intentions for this spectrum and what are your intentions today?

Mr. TUCKER. Senator, I would say that there has not been a classic bait and switch. In our company, as I said earlier, we are doing HDTV news in Seattle right now. The opportunity does exist within the law——

The CHAIRMAN. May I interrupt? Your company may be doing it, but there is a very small number of companies that are. I think you are representing NAB here, not just your company.

Mr. TUCKER. Yes, sir, that is correct.

The CHAIRMAN. OK.

Mr. TUCKER. I think that there is a lot of experimentation going on with this spectrum right now, and that was authorized under the statute, as far as I know. The ability to do an experiment——

The CHAIRMAN. May I interrupt one more time? 182 DTV stations, or 11.3 percent of the nearly 1,600 TV broadcast stations, have been made DTV capable.

Mr. TUCKER. Yes, sir, and that is ahead of schedule.

The CHAIRMAN. That is ahead of schedule?

Mr. TUCKER. That is ahead of schedule.

The CHAIRMAN. We are going to reach 85 percent of the homes in America by the year 2006?

Mr. TUCKER. We have 1,100 more stations to come on the air.

The CHAIRMAN. We are going to reach 85 percent of the homes in America by 2006, you are going to comply with the legislation? There is no one in America that believes that, Mr. Tucker.

Mr. TUCKER. Mr. Chairman, I think that date was condensed from the Congress. Originally, the expectations and the projections that the broadcasters looked at, was for that transition to take as long as possibly 2015.

The CHAIRMAN. I think you will find congressional testimony to contradict your statement, sir, by the broadcasters.

Mr. TUCKER. Well, the FCC does not have the same date, sir.

The CHAIRMAN. I will be glad to provide it with you. Go ahead. Please finish your response.

Mr. TUCKER. There is no question that we are encountering some problems with construction permits and FCC applications being held up, and that there probably will be some delays. There is an opportunity to go and ask for extensions for those stations that encounter those kinds of delays. But we expect the industry to be on track. As I said earlier——

The CHAIRMAN. Track to do what?

Mr. TUCKER. On track to be broadcasting by May of 2002.

The CHAIRMAN. What percentage of the industry?

Mr. TUCKER. As far as I know, sir, all broadcasters are projecting to try to be up on the air broadcasting by that date.

The CHAIRMAN. I was not asking you about try to be. I was asking you about when they would be. I mean, we are going to have further testimony here that there is not a snowball's chance in Gila Bend, Arizona that that is going to happen by the year 2002, Mr. Tucker, by objective consumer groups. You are telling me that 85 percent of the American homes in America will be receiving high density television by the year 2002, which means then that you would be prepared to give back the analog channels that you have also been keeping. Is that correct?

Mr. TUCKER. Well, there are two different questions here, sir. We already reach 67 percent of the country with the stations that are on the air. The incremental of the other 18 percent I think is doable by 2002. But the 85 percent set penetration means that the consumer has to have access to that programming.

Right now we have 70 percent of the homes that cannot get it because of cable's roadblocks, so we do have some problems in those areas. We also do not have interoperable receivers or digital television sets that enable the consumer to get HD television at this particular point in time.

The CHAIRMAN. Cable operators, Mr. Tucker, did not ask for $70 billion worth of spectrum, you did. Therefore, I am interested in hearing what you intend to do to increase the amount of programming that would give the consumers a reason to go out and purchase digital television sets.

Mr. TUCKER. Mr. Chairman, we already are doing more programming. As I said, we have over 1,000 hours of programming on the air right now. We are encouraging the networks to do more. Mr. Sagansky is right; in HD, high profile sports and theatricals are probably the best HD example. I think that there is a lot of innovation going on in the broadcasting industry to provide different digital services in multicasting and in HD combinations.

The CHAIRMAN. When do you expect the analog channels to be given back?

Mr. TUCKER. As soon as we can reach 85 percent set penetration, sir.

The CHAIRMAN. Which is when?

Mr. TUCKER. I think the consumer is going to determine that.

The CHAIRMAN. Mr. Willner, one of the factors delaying the DTV transition is the inability of the content owners and transmission providers to resolve copyright issues. Do you know anything about that?

Mr. WILLNER. A little bit, Mr. Chairman. I think that we have to walk the line of protecting the consumer's right to be able to do some time shifting and watch a program versus the copyright protection of the intellectual property owners. But I think there are probably technological solutions that would allow us to be able to transmit those programs.

The CHAIRMAN. According to the FCC, where cable companies face competition from a telephone company offering cable service, the price of cable measured on a per channel basis is 35 percent lower than where cable faces no competition. Please explain to me how in these communities with no competition, consumers are truly getting the value they deserve.

Mr. WILLNER. Mr. Chairman, I do not know where there is a 35 percent difference. We have competition, we have markets where we have wireline overbuilders. Our rates are no lower in those communities than they are anywhere else. Our belief is to roll out technology that will compete against satellite, which is a very significant competitor in our industry; they have gone from nowhere to 15 percent penetration nationwide in the last 5 years, and quite frankly, I do not know where the price differences are. They do not exist in my company.

The CHAIRMAN. Senator Wyden.

Senator WYDEN. Thank you, Mr. Chairman.

Gentlemen, I want this issue addressed using marketplace forces. And to me, the real question for this morning's hearing is what is it going to take to unleash those kind of marketplace forces. It seems to me if the advantages are significant here, consumers would actually go out and start demanding this product, and suffice it to say, there aren't a lot of folks marching the streets today asking for DTV-capable sets. The majority of sets sold last year were traditional sets.

What I would like to do for my questions is to go down the row and have each of you state what the real advantages are here, so that the public gets a sense of what they are going to win if this issue is resolved, and why are marketplace forces not making it possible to get those benefits.

Let's start with you, and we will go right down the end.

Mr. SAGANSKY. Thank you. You know, I want to just go back and look at what drove cable acceptance 20 years ago. There were only two factors. One was better pictures, and two was better content. And when I say better content, I meant content that was proprietary toward that platform, that is HBO, Discovery Channel, things that you could not get over the air. And we maintain that it is going to be the exact same factors that drive the acceptance of DTV, that is better pictures, high definition pictures, standard definition television, as well as better content.

And when I say better content, there are a lot of broadcasters like us, like public television, that do not feel that the best use of the spectrum is a single stream of programming, but multiple channels, channels that will only be able to be received on digital television. So those two factors are going to drive consumer acceptance.

Then the question is, how does the consumer get it? Right now you can go out and buy a television. You cannot hook it into digital cable, because there are no accepted standards for digital cable hookup. So that digital television is useless.

Then the FCC comes along and says, "hey look, you cannot start, you cannot ask for must-carry on your digital until your analog is turned off." And my question there is, who is possibly going to buy a television set if they cannot get the digital, if they cannot get our digital signals over the air? Why would you buy a set? You will not.

So there has to be some sort of must-carry for our digital signal so that we can encourage people to go out and buy a television set.

Senator WYDEN. OK. Let us hear from Fisher Broadcasting.

Mr. TUCKER. Senator, thank you. First of all, with 70 percent of our viewers not capable of getting our signals that are sent through cable at this particular point in time, we do need cable cooperation to drive the marketplace. It is an unfortunate situation but that is just the reality that we live in.

I think the fact that we deliver unique public interest programming, local news and local public service, is still the basis and the foundation for what local broadcasters do. I think that that has been recognized by the consumer over and over again to be a high-demand product.

Senator WYDEN. Let me ask you another marketplace question. Your second recommendation—you all made three recommendations, the broadcasters today—involved this matter of the digital tuner. Now clearly that could involve cost to consumers. Do you have a plan so that there would be a volume of scale so as to not hit the consumers with a significant cost there?

Mr. TUCKER. Senator, only the experience that I think that we all recognize with any consumer product's introduction, that once we get to mass production, that that particular cost will become de minimus over time, the more sets that are rolled out.

Senator WYDEN. So what does it cost at the beginning and under your theory, what does it go down to?

Mr. TUCKER. I think in the end there will not be a significant cost factor at all.

Senator WYDEN. So your three proposals as the broadcasters envisage would not involve significant cost to consumers?

Mr. TUCKER. I do not think so, no, sir.

Senator WYDEN. OK.

Mr. Willner.

Mr. WILLNER. Well, Senator, let me start by telling you a story that happened about 20 years ago, because I think it is significant here. I was in my office as we were building our cable system in Clearwater, Florida, and a fellow by the name of Bud Paxson came in with an idea in his mind that he wanted to start a very different type of television program that was not offered by anybody else in the cable industry. And in 15 minutes he had worked out a carriage arrangement with me which started the Home Shopping Network.

The fact of the matter is that the broadcasters are trying to make it into a government mandate about whether or not cable has to carry multiple signals or whatever it is that they want to do to in order to make money delivering their signals. This debate is about whether or not the cable operators have the consumer's best interests in mind when they do not want to broadcast duplicative signals, which will take away bandwidth from high speed data access, from voice telephony alternatives to local phone company, and from video on demand products, which are all very bandwidth intensive.

I do not think that is the best use of our bandwidth or in our consumers' interests. We have markets now where Insight has launched our digital products, and these are real digital products going out, with interactive capability, and we have up to 30 percent penetration within 1 year of customers taking digital cable. So there is a market-oriented way of rolling out digital television.

I am not sure I understand how the broadcasters are going to do it, but I do know how the cable operators are doing it.

Senator WYDEN. You have to wonder why we need this hearing. I mean, in one sense, if markets are working except in rural areas—Senator Smith and I have a rural State, Senator Rockefeller, Senator Dorgan, we have got some markets where because of the distance, rural communities, small numbers of people, I can see why this would be slow to kick in. But it is not kicking in anywhere, so we are going to have to look with you to find some additional incentives.

I am particularly interested in the cost of the broadcaster package. All of you at Fisher, I think have been very community-minded on a number of instances, and without getting into it, I appreciate your supporting my stand by your ad proposal, the bipartisan proposal on campaign finance, not the topic for today, but we will look at the cost of your proposal as well, because there is a reason that consumers are not rushing to this product and that is that up to this point, the markets have not worked very well.

And if you as the broadcasters have a proposal that will drive the costs down here, that is certainly worth looking at.

I thank you, Mr. Chairman.

The CHAIRMAN. Senator Burns, before I go to you.

Mr. Tucker, on September 17th, 1997, at a hearing before this Committee, Mr. Robert Decherd, Chairman, President and CEO of A.H. Belo Corporation on behalf of the National Association of Broadcasters, said quote: "Broadcasters have made a compact with the Congress concerning high definition television. We will meet our commitments. This compact is essential to our future since it insures the long-term viability of free over-the-air television that American citizens have enjoyed. Now is the time to move to HDTV."

Senator Burns.

Senator BURNS. Thank you, Mr. Chairman. To the list of numerous difficulties through this whole thing, I have a couple of questions. In your estimate, Mr. Tucker, is there something that the FCC can do that would facilitate or to make this transition go a little bit faster and maybe smoother? Any recommendations that you would make to the FCC or to this Committee that the FCC could do?

Mr. TUCKER. Yes, sir, I think I have already asked for it. Must-carry and receiver set standards.

Senator BURNS. What do you believe is the single largest impediment right now in this transition, if you had to deal with just one? And I would welcome Mr. Sagansky's comment on that also.

Mr. TUCKER. I would say must-carry.

Mr. SAGANSKY. I would also say must-carry. It is the single biggest thing because if we are getting the programming out there and people can receive it, then they will see the difference. It is an unbelievable difference. It is more content and it is better content. It is clearer content. That is the thing that drove us to say hey, we want to spend the $7.5 billion to convert to digital broadcasting to begin with.

We think there is a huge difference for the consumer.

Senator BURNS. Give me an idea of the transition being made by the smaller and medium size and smaller markets. Where are they now?

Mr. TUCKER. Senator, they are due to be on the air by May of 2002. There is no question that there may be some delays in those markets. They also expected the transition to be further along. They expected sets to be in place and some part of must-carry to have occurred so that there was an expectation that the consumer would get their product.

I think that if the FCC were to expand the ability for them to ask for a delay based on economic hardship, the smaller markets might do that. But that does not speak to the majority of the American public. The larger markets have the ability up to about market 75 to deliver to close to 80 percent of the country.

Senator BURNS. I guess one would always worry about where he lives and how that affects you personally. I am in what we would call a small market. Of course, we have a tiny market in Montana where we are riding dry there, but what about the Billings, Montanas and the Casper, Wyomings, and for the most part, some areas in Alaska?

Mr. TUCKER. Senator, we have got a television station in Idaho Falls, Idaho, a 163rd market. We are going to be on the air by 2002. We have small terrestrial satellite stations in Coos Bay, Oregon and Tri Cities, Washington, and Lewiston, Idaho. Those will all be on the air as well, but we have the financial capability to do that.

Single market small station owners are going to run into strong economic hardships, there is no questions about it. It will be better for us to be pioneers as broadcasters in the digital era to kind of facilitate for them some of the learning that we have to push the prices down for transmitters and to see sets start to come out so that the consumer is attracted to it, and move the demand forward.

Senator BURNS. Now the next question gets on the other end on the consumer part. Now I was out at the consumer electronics show and last year when they were asking $4,000 bucks for a receiver of digital television and the screen and the whole thing, it was costing around $4,000 bucks. This year, and I would agree with you, that cost had almost been cut in half this year they tell us.

But still, as Scotch as I am, I am not ready to give $2,000 for a television set, and I am wondering what the attitudes are in Idaho Falls; are there more like me or are they going to stand in line to order these television sets?

Mr. TUCKER. I think they are far more like you. I can tell you in Seattle, we only have about 10,000 HDTV sets in the market and we have all four major broadcasters that are up with HDTV. Of those 10,000 sets in the Seattle market, only about 1,500 receive over-the-air television at this particular time.

Senator BURNS. So Mr. Sagansky, you are getting nervous here.

Mr. TUCKER. It is going to take a while, Senator, I mean it really is.

Mr. SAGANSKY. Look, I do not think anybody, Senator, is going to go out and buy a $2,000 set if once they plug in their cable there is no programming. So you are right. I mean, it is not only the cost of the set and nobody is buying the sets because there is no interoperability standards for cable, and there is no must-carry so there is nothing on the cable to watch that is free, it is all just digital cable channels. So it is a huge problem.

You know, if you go back to the 1992 Act that mandated must-carry, it was one of the most successful Acts in this history of this country in promoting diversity. You know, it gave rise to the WB, UPN, Telemundo, Univision, PAX, and a whole bunch of minority and religious broadcasters that could never be on the air before and get through to the consumer, because it mandated cable carriage. That is what we are asking today. We want to program for these digital consumers but it has to get through to the 70 percent of the consumers that have cable and satellite.

Mr. WILLNER. Senator, could I respond to that?

Senator BURNS. Yes.

Mr. WILLNER. I would just like to say that first of all, when the government decided that it was in the public interest to deliver to the broadcasters additional spectrum to the tune of about $70 billion, they did so with high definition television as being the reason for that.

To the extent that broadcasters now want to multicast and deliver different types of programming to create different types of revenue streams, I do not believe it is in the interest of the government to determine whether or not they have the right to do that over any privately-built network that has been put in place around the country.

The fact of the matter is that the biggest impediment to the delivery of digital television today over the broadcast network is the broadcasters' lack of a business plan and lack of content over that platform. To the extent they want to multicast, that is fine. Come talk to us and let us work out a business arrangement that makes some sense. You know, must-carry was always originally intended to protect the local broadcasters' economic well being, to deliver news and to deliver weather, and local information in communities like New York City and like Idaho Falls. The problem is that we are now discussing delivering to them bandwidth that allows them to simply make money. It is not really necessarily in the public interest.

Senator BURNS. We know where the must-carry language came from. Thank you, Mr. Chairman.

The CHAIRMAN. Senator Dorgan.

STATEMENT OF HON. BYRON L. DORGAN, U.S. SENATOR FROM NORTH DAKOTA

Senator DORGAN. Mr. Chairman, thank you very much.

We are speaking here about HDTV and digital in some ways as if they are one and the same, and they are not. I am just trying to understand the circumstances of all this.

Let me read something to you and see whether you think this is fair, and I guess what I am going to read describes part of the responsibility for the delay with respect to virtually everyone who is testifying and will testify.

Broadcasters are frustrated with equipment manufacturers, as I understand it, because that has not gone very well and they are not producing a lot of sets that are less expensive.

The equipment manufacturers are upset with the broadcasters because they say the broadcasters challenge the broadcast standard chosen by the FCC, which I believe the broadcasters now support.

Equipment manufacturers and the cable operators haven't gotten together on interoperability standards to make sure purchasers who purchase the sets will be able to use the sets.

It seems to me like you have a little blame laying at everyone's feet, and you probably could make the case that the FCC has not been particularly aggressive in providing leadership here. So, virtually everyone has a share of this, and the question today is, what do we do next?

The Chairman makes the point that we have got spectrum out there for the purposes of conversion and he makes the point that it does not look like conversion is moving along with any great speed.

Let me ask Mr. Willner, you testified that, in opposition to must-carry, you don't want the government involving itself in these issues. Let me ask you, is your opposition a philosophical one or a physical capacity one?

Mr. WILLNER. Well, it is both. I mean, the fact of the matter is that there has been a lot of talk about a 750 MHz cable system having unlimited capacity and therefore, you might as well just give the broadcasters another channel. The reality of a 750 MHz cable system today is that 550 MHz of that system are allocated to the old bandwidth hog-analog television signals.

The reason for that is because we are not trying to force digital down anybody's throats. We are trying to offer it as an optional tier so that we can create new revenue streams that will alleviate the pressures on the cable industry to raise basic rates, which I know is a concern of this Committee as well.

And optional digital tiers are really the right answer in a world where our programming costs are going up 15 to 20 percent each year, and we are trying to keep our rate increases down at the rate of inflation.

I would also like to correct one statement that you made before about the consumer electronics and cable industry not having an agreement. We do have an agreement on interoperability, it was filed with the FCC last year, and right now there are television manufacturers and box manufacturers who are beginning to make equipment that will allow cable and television sets to interact with each other.

Senator DORGAN. Mr. Willner, if you were to purchase a television this afternoon in that case, then what would you purchase that would give you the feeling you could hook it up and achieve the signal you wanted off your cable system? Does that exist?

Mr. WILLNER. Well, I think you could buy a high definition television set and determine what type of a tuner you are going to buy later on, because the pieces come in components.

Senator DORGAN. But that is my point, is it not? And we are making the point that consumers are not out there able to buy at this point with some knowledge that if they come home and hook it up to their cable box, they are going to be able to advantage a signal that justifies the price they paid for that set. Is that not it?

Mr. WILLNER. In Columbus, Ohio, 30 percent of our market is taking digital, they are buying our digital product at an incremental revenue of $22 per month. They are volunteering to pay it. They do not have to buy anything. We are delivering the equipment to them as part of that package.

Senator DORGAN. Let me ask a question of Mr. Sagansky. The auction for the spectrum occupied by channels 60 to 69 has been delayed a number of times, and I believe your company has been describing an interest in having that happen.

What happens to your non-cable viewers if you were to leave that spectrum without a place to go? I did not quite understand the answer you gave.

Mr. SAGANSKY. The point is that we are willing to leave that spectrum. But we are going to lose our over-the-air viewers, unless they go out and buy a digital set. And so we are going to lose, you know, in some markets as much as 40 or 50 percent of our viewers.

But you know, in the interest of clearing the band early so it's used for advanced wireless, 3G wireless, we're trying to devise a plan that will help—you know, the people that are bidding on this so that they know that there is an agreement that we will leave the band.

Senator DORGAN. Now let me ask the three of you a question. The Chairman has been asking questions about when will we reach the goal that we had intended, the 85 percent. It seems to me like we are crawling toward that finish line without great progress. And it is quite clear to me from the testimony the reason for that. We have several different interests that are moving around trying to prevent certain things from happening.

Do any of you have any hope that we are going to reach an 85 percent penetration of consumer sets at any point in the reasonable future here? I mean, can you give us any estimates? I do not think the Chairman got an answer on that, but we are talking about the 85 percent penetration with customers having sets that are able to access the digital signal. When is that going to happen?

Mr. Willner.

Mr. WILLNER. I think it is going to take a change in the way the broadcasters have been viewing whose responsibility it is to deliver content over this digital bandwidth that you have awarded to them.

Senator DORGAN. So you don't know.

Mr. Tucker.

Mr. TUCKER. Senator, first of all I would like to say that we did get the transmission standard taken care of in January, so at least we are in agreement as a broadcast industry that 8VSP is the standard that we are going to transmit with. And there was some confusion in the marketplace about that.

I would say that when we do not have access to 70 percent of our viewers, no matter how good the pictures are that we put up, if they are not passed through by the cable system, we are not going to drive those sets into the household. So until we get that, it is going to be some time.

Senator DORGAN. Mr. Sagansky.

Mr. SAGANSKY. Yes. All I can say is from the broadcasting standpoint, I know we will be ready with the pictures. We will be broadcasting digitally. Whether anybody out there is going to be able to receive it, I do not know. I think it is going to take the FCC and Congress together to say, hey look, we have got to make this happen, for a while you are going to have to carry those digital signals until we are fully penetrated, and then you can get rid of the analog.

And that is what it is going to take, because unless people can get it, they are not going to buy the sets.

Senator DORGAN. Well, Mr. Chairman, thank you. I did not ask questions about the rural areas because we know that is always going to be the tail on the dog here, and even if you get all of this solved and you begin to get density with respect to sets that can receive the digital signal, what is going to happen to Montana and Alaska and North Dakota? I know exactly what is going to happen here, and I will ask question about that at some later date. But I think the Chairman has expressed the frustration for many of us on the Committee that we need to get these issues resolved. I am real pleased that he is holding these hearings.

The CHAIRMAN. Thank you, Senator Dorgan.

Senator Stevens.

Senator STEVENS. Thank you, Mr. Chairman.

Until we decided to, and Congress decided it, to auction spectrum, those who made application to the FCC for licenses, you got it free. And when we got to the point of trying to see if we could move into this new technology, we decided we would give broadcasters an area to convert to and to release the analog spectrum they had in the—if they did, when they did.

I am informed now and I want to make sure that I understand this right, that with the 6 MHz of spectrum a broadcaster can either, it is an either or, broadcast one high definition TV signal or it could broadcast up to 6 channels under current technology of the digital. And if I understand this right, your testimony, Mr. Sagansky, there are 33 million analog TV sets and 26,000 digital tuners in the country so far.

Mr. SAGANSKY. And that was just sold this year, sir.

Senator STEVENS. This year?

Mr. SAGANSKY. Yes.

Senator STEVENS. I am like Senator Burns. I have enough Scotch in me—I said that once by the way, Mr. Chairman, on the floor, and someone in the House said I was admitting that I was half drunk.

[Laughter.]

But I'm half Scotch, and even a half Scot doesn't buy—I don't buy digital before the signals are there. So the really great problem I have here is what are we going to do to give an incentive to people to start buying the sets that will receive these digital or the high definition TV that all of you have indicated is out there sometime in the future?

Don't we still have the cart before the horse? I mean, the horse really is, the driving force is the consumer, and the consumer does not have the sets.

Mr. Willner, you do not have either, do you?

Mr. WILLNER. No. We deliver our digital signals over the cable system and convert them in the home so that an analog television set can pick them up. The picture quality is very very high on that.

Senator STEVENS. I understand that. But you are not racing to get digital TV sets out there because you do not care, do you?

Mr. WILLNER. No, I do not.

Senator STEVENS. Right. And these guys do, because they are sort of mandated to make the conversion that does not affect you at all.

Mr. WILLNER. It does not affect me at all except to the extent that they try to confiscate our services.

Senator STEVENS. So I do not really think you ought to be the one that is pulling the cart either, because the problem is, how do we deal with these older TVs, and I am going back to the consumer. When you look at my State, more than half of the people who receive signals in my State receive public television signals. As a matter of fact, I think we are the only State that rebroadcasts signals at State expense to get to areas where they do not have reception otherwise. We put it up on there and they bring it down off the satellite.

Now when you look at this, what is going to happen to the consumer? What is going to happen to the college classes, to the basic classes even in the schools which are now going into what, teleeducation, or the medical facilities, telemedicine. What is going to happen to them in terms of this problem, in terms of this conversion from analog to digital as far as the system is concerned?

Mr. Sagansky, you indicate that you believe that the cable systems ought to broadcast all of the multicast signals you can put within this spectrum, right?

Mr. SAGANSKY. Within our 6 MHz that we have been allocated, right.

Senator STEVENS. And currently that would be up to 6 channels, right?

Mr. SAGANSKY. 5 or 6 channels, yes.

Senator STEVENS. And I am informed it is not too far away that we are going to start splitting that down and you will be able to—just like radio—to be sending more than 6 over that spectrum. Do you think as that technology improves, they should be required to carry whatever you can produce to carry over the 6 MHz?

Mr. SAGANSKY. I think as long as we are just broadcasting within that 6 MHz that whether we choose to do a high definition or multiple channels, or all news, or we appeal to some sort of ethnic group in each of our markets because we feel that is in the public interest and it is a business for us, whatever we choose to broadcast I think should be carried by the cable company.

It is no more bandwidth for them whether they carry an HD signal or these 6 channels. It is still only 6 MHz.

Senator STEVENS. What are you going to do about the people that still have the analog sets?

Mr. SAGANSKY. Well, those people will continue to get an analog signal until we reach this 85 percent of penetration of digital.

Senator STEVENS. So what you are saying now is you are going to require them, you want the cable people to carry both, do you not?

Mr. SAGANSKY. We do, but our plan is, what we would like them to do is as they build out their systems, as they put in their digital boxes, that is when they start carrying our digital signals.

Senator STEVENS. Well, what is the incentive then to me to ever buy a digital set?

Mr. SAGANSKY. Just only two things. One, because the picture quality, whether it is high definition or these 6 channels, which are all in standard definition, which are much better than what you are seeing now on your analog set.

Senator STEVENS. Yes, but you see, the difference is once you get must-carry it doesn't make any difference, because with his signal, he straightens that up. So my set—you can be broadcasting in analog, but put it through the must-carry, and it is coming to me in digital form, as I understand it. Am I wrong, Mr. Willner?

Mr. WILLNER. No, you are correct. And there are also numerous agreements already in place between broadcasters and cable operators to retransmit digital signals as a part of retransmission consent. So all this is is the marketplace at work. And what is really going on here is that there are broadcasters who are asking the government to do their job in negotiating deals with cable operators.

Senator STEVENS. But if I am informed right, and I am getting to the end of my time, but if I am informed right, most of the entities have agreed to carry at least some of the broadcast signals, not all of them, but some of them. The FCC decided there was no requirement to carry more than one, right?

Mr. WILLNER. That's right, just one, which could be only 1 MHz.

Senator STEVENS. But if I am informed right, Time Warner and others have agreed that they are going to carry more than one, but they are not going to be mandated to carry all.

Mr. SAGANSKY. They have only agreed with the big operators like the News Corp. and Viacom. They have market power.

Senator STEVENS. I am talking about public stations now. They have agreed to carry the public stations but have not agreed to carry you, right?

Mr. SAGANSKY. Us and thousands of other little broadcasters. Virtually no one I know other than the huge conglomerates that they need to carry because they have so much market clout.

Senator STEVENS. Well, I will tell you, again, I go right back to the question of the sets. Until we get some incentive to the Scotch consumers to buy additional tuners, I do not know how we can keep up with the timeframe that was in the original Act. You disagree with that, Mr. Willner, right? You think we should not change the timeframe of the original Act?

Mr. WILLNER. I do not think that the course that we are taking is going to change anything. I think that the broadcasters have to determine what they want to do with that spectrum and go ahead and do it, and that is what will change the course of the transition to digital.

Senator STEVENS. But I have a problem with that because even if they do what we thought they were going to do, if John Q. Citizen does not have a digital set, it does not make any difference.

Mr. WILLNER. Well, we will be able to deliver high definition television signals over the cable system.

Senator STEVENS. Yes, you will, but they will be delivering to people over the air in areas where they do not have cable, and they will be delivering a digital signal and I have an analog set out in my place out in Girdwood, and I have to tell you, I am not going to get any signal at all, am I?

Mr. WILLNER. If they have turned off the analog stations, that is correct.

Senator STEVENS. Right. That, Mr. Chairman, is the worst problem. If we do not do this right, we are going to isolate the rural consumer and my State is all rural, so guys, we had better get together here pretty soon because I want to be on board when the crash comes, and I have to be on board at the takeoff. I do not think this is going to work unless we change that law in some way.

Thank you, Mr. Chairman.

The CHAIRMAN. Senator Brownback.

STATEMENT OF HON. SAM BROWNBACK, U.S. SENATOR FROM KANSAS

Senator BROWNBACK. Thank you, Mr. Chairman, and thank you to the panel for being here.

I think you can sense all of our frustration in dealing with this and I hope it spurs you on to doing a lot more to getting DTV deployed and moving on forward.

I come from a rural State as well and one of the things that I am concerned and interested about that is taking place is the digital divide and the lack of high speed Internet access to rural areas. One of the ways to be able to broach through that is through wireless, which is going to go through a number of the channels that you folks are occupying right now, because a number of these advanced services are anticipating being able to use some of those various channels. If the DTV transition keeps delaying longer and longer, as the digital divide on high speed Internet access gets wider, there is going to be more pressure on Congress to act, and there is going to be another sector that is going to be pushing on this aggressively.

I hope if the hearing does not do anything else, I hope it spurs you to say this is not about delaying and trying to occupy this bit of real estate for a longer period of time, and I hope it spurs the cable industry as well, into saying, "what can we do to work with this." Because otherwise, there is going to be some pressure put on Congress that you probably do not want to see take place.

I work a lot with the wireless industry and they are bumping up against these spectrum caps and they are needing more spectrum now, and they would do much more deployment of the advanced and 3G services but for the lack of spectrum. I think to date they have not pushed and screamed and hollered too much, but they are set to and they are getting ready to, and that is going to add pressure in this field.

You hold the answer and the blockage to a couple of our problems. One is this digital divide in some of the rural areas that we are seeing take place on a massive scale, and we are looking toward this spectrum as a potential part of the solution.

A second is the advanced wireless services that a number of us are looking at as great potential for a whole host of activities, and they are dependent upon this spectrum becoming available and useful.

My point to all three of you would be that if you do not do something, probably something is going to happen and move forward, and that pressure is going to build in an ever aggressive amount. I have had a frustration that it appears as if in some sectors there either has not been enough impetus pushed to cause, that says this date is a hard and fast date, it is one we want to hit, or the people there have felt like it is just not imperative that these things start really moving aggressively.

The pressure is building strong for this to take place.

Mr. Chairman, I have a full statement I want to introduce into the record, if I could at this time.

[The prepared statement of Senator Brownback follows:]

PREPARED STATEMENT OF HON. SAM BROWNBACK, U.S. SENATOR FROM KANSAS

Mr. Chairman, thank you for convening the Committee to once again review the transition to digital television. Unfortunately, in the more than 2 years since our last review of this issue, it appears that long-standing and some new disagreements between industry participants continue to stall this transition.

Since our most recent DTV hearing in 1998, the broadcast, cable TV, and consumer electronics industries have failed to resolve issues associated with interoperability, content protection and fair use, and HDTV versus multicasting. In the interim, an internal debate among broadcasters regarding the DTV broadcast standard itself did little to speed up the process. Most recently, the Federal Communications Commission's January 22nd decision rejecting dual analog and digital must-carry, as well as permitting cable operators to modify DTV signals they do carry, have unfortunately failed to prompt industry to renew their marketplace efforts instead of looking for resolution at the Commission or in Congress.

I am eager to see the development of a transitional process that embraces the marketplace and the freedoms a market-based transition affords the affected industries. I, for one, cringe at the idea of a Federal bureaucracy, and Congress above all, stepping in to set standards and regulate this transition.

While these outstanding issues remain to be resolved, I feel compelled to spend the balance of my time addressing an equally important and interdependent issue: how the continued failure to resolve issues slowing the DTV transition is threatening to deprive the American public of other much-needed services.

Wireless broadband Internet access raises the prospect of helping to bridge the digital divide by providing many of my constituents, and rural communities across the nation, with the services and content broadband Internet access makes possible. While the wireless industry has an abundance of desire to deploy 3G services, the next generation of wireless network technology, they require additional spectrum resources to do so. The DTV transition is not simply about providing the broadcast industry, already the beneficiaries of tens of billions of dollars worth of free spectrum, with new capabilities and revenue streams. It was also intended to free up spectrum resources for new services such as 3G. The stagnant nature of the DTV transition is reducing the likelihood of achieving these twin goals.

While the Committee reviews many of the same issues we focused on in 1998, I hope we all recognize that this transition is starting to become less about the new services digital television makes possible, and more about the opportunities the transition's continued stagnation will deny the public. I think we can all agree that isn't an outcome we're interested in.

There's another point I want to raise: as the industry makes use of the enormous opportunities that digital television provides, it is timely and appropriate to consider the obligations incumbent upon them to serve the "public interest, convenience, and necessity." Numerous studies have documented the decline in standards in some parts of the broadcast industry, resulting in the high levels of violence and steady rise in vulgarity that characterizes so many programs. We can do better. In addition to the adoption of a voluntary code of conduct on the part of the industry, we need to have an honest discussion about the nature and extent of broadcasters' public interest responsibilities.

I urge the industries driving this process to redouble their efforts to resolve these outstanding issues. I look forward to a transitional process that not only succeeds in deploying the exciting capabilities of DTV to our homes, but one that also promotes the deployment of other vital services to the public.

Thank you Mr. Chairman.

The CHAIRMAN. Mr. Cleland.

STATEMENT OF HON. MAX CLELAND, U.S. SENATOR FROM GEORGIA

Senator CLELAND. Thank you very much, Mr. Chairman. May I say that I can remember in my little hometown of Lithonia, Georgia when I first actually saw a television. It was 1950, I was in the third grade, and I can remember the first TV program I ever saw was the "Lone Ranger," and I grew up thinking I was the Lone Ranger. And television had a powerful impact on my life, and later on I identified with other cowboy heroes like Roy Rogers and others. It was unbelievable to my mother how much time I actually spent in front of the television.

I have been fascinated anticipating the arrival of digital television, DTV. To ease the transition to DTV, Congress allowed 10 years for its completion. However, we are not on schedule to meet the 2006 deadline, and Roy Rogers and the Lone Ranger are dead.

There are forces in the marketplace which we cannot control, like the demand for and the price of the sets, which I understand can be upwards of $5,000. I think my first television secondhand cost $50. In 1950 terms that was a lot of money, but $5,000 is still a lot of money in today's terminology.

But I think we can, on the other hand, encourage all of our relevant parties to work together and settle the disputes that are standing in the way of reaching the goal.

I would just like to recognize that some industries are transmitting digital signals currently. Consumers are able to receive a digital broadcast from cable and satellite companies. Additionally, virtually all of the top 30 market broadcasters and many of the remaining markets are transmitting a digital signal for those households able to receive it. I applaud these industries for their work.

The United States has been the leader in broadcasting and entertainment. I hope that the problems we have encountered so far will not stifle our leadership. We live in an age, as we all know, where things happen in Internet time. What is it Bill Gates says, "business at the speed of thought." And I think we probably need to put a lot more thought now into how digital TV comes into the marketplace.

I think we have got to be flexible, I think we have got to understand that compromise might be what we need to do, and make sure that this state-of-the-art product is not overshadowed by technology that we have not even given a thought to.

I would just like to ask all of the panelists here, what are your suggestions for us here as a Committee? What do you think we should do? Do you think the 2006 deadline is too soon, too quick? I mean, putting a deadline on technology and its implementation is like putting a deadline on withdrawing the troops from Bosnia. It may be completely irrelevant.

Mr. Sagansky, any ideas?

Mr. SAGANSKY. A couple of things. First of all, the deadline of 2002, I think, is going to be very very hard to hit. For a lot of broadcasters, they will hit it. But half the broadcasters have not even gotten their DTV construction permit. They have not gotten the permit to build the digital station.

So here we are 14 months away. If you have not gotten it now, you are not going to be ready by 2002. You cannot build them that fast. There is not the availability of the equipment that can come that fast. So that is one of the big problems.

And then, the consumer has to have a plug-and-play TV set, just as easy as you go out and buy an analog set. You go home, you hook it up to your cable, you are on. You do not have to think about anything more. And right now there is no digital cable standard. You cannot plug it into cable. There is no standard. Then you have to go out and buy a tuner, so you have to buy a tuner and a monitor. And then you have to go and put a rabbit—you know, you have to go and put an antenna on your roof. I mean, this is what you had to do when you were watching the "Lone Ranger."

Senator CLELAND. I still have the rabbit ears in my basement.

Mr. SAGANSKY. You are going to need it again.

That is obviously a very critical component to this, you know, the plug and play.

Finally, once you plug it in, you get all that stuff done, is there going to be anything there worth seeing? Right now, if the cable companies do not have to carry the programming, our free programming that we want to produce and put over, then what is the point? You are not going to get clear images. You will not even see our programming at all, because they do not have to carry it.

So those are the three things that have to happen for this thing to work.

Senator CLELAND. You are with PAX-TV, right?

Mr. SAGANSKY. That is right.

Senator CLELAND. They are a family-friendly network, and I applaud you for that, and good luck to you in that regard.

Mr. Tucker, where are we on this? Is the 2006 deadline unrealistic or are we just dreaming here, or should we just forget that and let the market forces take us wherever we go?

Mr. TUCKER. Senator, I agree with that, with the three things that Mr. Sagansky mentions, and I would also say that the broadcasters as an industry will have stations, I expect them to have their stations up and running well before 2006.

But to get to 70 percent of our viewers, we need to get through the cable gatekeeper. And to incentivize the viewer to watch HDTV, they need to have access to that programming from their cable system. We are only talking about the same 6 MHz of spectrum for the cable company whether we are doing HD or standard definition television. I think that will be the biggest catalyst, to get must-carry taken care of. I think 2006 is going to be a very difficult deadline to meet.

Senator CLELAND. Thank you. I would agree with that.

Mr. Willner, comments, suggestions?

Mr. WILLNER. Senator, thank you. First of all, I would like to correct for the record that there is an interoperability agreement in place between the consumer electronics industry and the cable industry to manufacture television sets that will be able to plug and play onto a cable system. So that is already in place.

I would like to just repeat some of my earlier testimony that we are already achieving penetrations of digital television in some of our markets in excess of 30 percent and I would hope that by 2006, more than half, and maybe three-quarters of our subscribers, will voluntarily subscribe to some sort of digital services.

I would like to remind my broadcasting colleagues that all it takes is a business negotiation to carry signals other than the one that the government was interested in protecting. Analog must-carry is fine, we will carry the local signals.

But to the extent that broadcasters want a special right over cable networks like Oxygen, like National Geographic, for must-carry status in order to build a business out of it, I do not understand that. Those people go out and raise hundreds of millions of dollars of high-risk capital to invest in additional programming, and then they come to the cable industry after they have done that and they negotiate carriage agreements. There is no difference between a broadcaster and a cable network when it comes to that.

Senator CLELAND. Fascinating.

Mr. Sagansky, do you believe that broadcasters have a public interest requirement in exchange for the free spectrum received in the 1996 Act?

Mr. SAGANSKY. I absolutely do. I think we exercise that. The one thing I would like to take exception to is something that was just said, and that is, our content is free to the consumer. We do not charge them. Unlike digital cable, we do not charge them. It is free. It is advertiser supported.

Senator CLELAND. Thank you very much.

Mr. Chairman, my time has expired. Thank you very much.

The CHAIRMAN. Thank you.

Senator Fitzgerald.

STATEMENT OF HON. PETER G. FITZGERALD, U.S. SENATOR FROM ILLINOIS

Senator FITZGERALD. Thank you, Mr. Chairman, and thank all the witnesses for being here.

I wish I had been in the U.S. Senate back in the late 1990s when this law was passed that gave rise to these problems we now have. I can assure you, Senator McCain, I would have been battling against what you called at the time a big taxpayer giveaway. In fact, when I announced my candidacy for the U.S. Senate, I cited the giveaway of the digital broadcast spectrum as one of the types of corporate welfare giveaways that I would like to go to Washington to fight.

I understand, Mr. Chairman, you read into the record parts of Mr. Safire's op-ed from the *New York Times*. I would like to introduce that full article, as well as several editorials that I brought with me, into the record. I would like to have unanimous consent.

The CHAIRMAN. Without objection.

[The information referred to follows:]

[From The New York Times, March 27, 1997]

GIVING AWAY THE AIRWAVES

(By Bob Dole)

WASHINGTON—The Clinton Administration, Congress and the Federal Communications Commission are about to make the already difficult job of balancing the Federal budget harder than it has to be.

In just a few days, the F.C.C. is going to give away the first broadcast licenses for digital television to broadcasters for absolutely nothing. If the Government sold this new spectrum, it could be worth $12 billion to $70 billion.

The network news programs are proud of their commendable watchdog segments like "The Fleecing of America," "Reality Check" and "Your Money." But the networks and many newspapers that own TV stations have largely ignored their own fleecing of the taxpayers.

The broadcasters insist that they need these airwaves—on which they will duplicate their programming in digital—to make the transition to high-definition television. O.K., but why not pay a fair price?

Since 1993, wireless phone and direct-broadcast satellite companies have paid for airwaves to upgrade or offer new services. Just last year, the Government auctioned off licenses for lower-quality spectrum, raising a whopping $20 billion.

We don't give away trees to newspaper publishers. Why should we give away more airwaves to broadcasters? The airwaves are a natural resource. They do not belong to the broadcasters, phone companies or any other industry. They belong to the American people.

The stakes in this debate are high. The national debt now exceeds $5 trillion. This year, the budget deficit is expected to top $112 billion. Balancing the budget with the help of proceeds from the new digital spectrum could lower interest rates by 2 percentage points, reducing costs for home mortgages and student loans.

While the needs of less economically viable stations and those in rural areas should be considered, broadcasters should be expected to pay for additional spectrum. After all, broadcasters have a long history of paying top dollar for existing channels. And the new technology will allow them to cram other potentially lucrative services—additional TV stations and CD-quality radio broadcasts—into the same amount of spectrum that currently accommodates one TV signal.

The Clinton Administration and a majority of the F.C.C. commissioners insist that the transition to the digital spectrum will allow the Government to sell the old analog spectrum by 2002. Indeed, the projected money from this sale is used by President Clinton to claim a balanced budget by then. But few believe Mr. Clinton's budget is really balanced, and even fewer believe the transition to high-definition TV can be completed according to his timetable.

As it is, this mandated transition to digital television is going to cost taxpayers plenty. Consumers will find their current televisions rendered obsolete by digital broadcasts. Replacing all 222 million TV sets in the country could cost upward of $200 billion. That's pretty serious sticker shock for "free" broadcast television.

Before leaving the Senate, I secured a written commitment from the Congressional leadership and all five F.C.C. commissioners to prohibit the distribution of digital-TV licenses until Congress passed legislation concerning the use of these airwaves. Strangely, no new laws have been passed, and the resolve of Members of Congress has melted (with the exception of Senator John McCain). And despite their commitment, F.C.C. commissioners plan to proceed with the giveaway.

Given recent developments, what's the rush? Broadcasters are scrambling to meet the Government's timetable. President Clinton belatedly proposed that in exchange for the licenses, broadcasters provide free time to political candidates. Still others propose trading licenses for a stronger TV-ratings system. At the very least, the next step should be to let the free market work and delay the giveaway until politicians and regulators get their priorities straight.

Taxpayers should demand better from the President, Congress, the F.C.C. and the broadcasters. After all, we're talking about billions of dollars—and that's your money.

Bob Dole, the former Senate majority leader, was the 1996 Republican Presidential candidate.

[From the St. Petersburg Times (Florida), July 18, 1997]

HOLD BROADCASTERS TO THE DEADLINE

You don't need high-definition television to see a naked display of corporate greed.

In April, the Federal Government loaned the nation's 1,600 TV stations a second broadcast channel. It was a giveaway worth billions—a legitimate public investment, according to the welfare queens of Wall Street, to ease the costly transition to cinema-quality TV.

Now broadcasters are dropping their end of the bargain. The TV industry wants to retain its new digital channels, while squeezing more life (and profits) from existing analog channels. Those channels were supposed to be returned by 2006 and auctioned by the government. Congress should remain firm. Any delay would further fleece taxpayers and send the wrong message.

Two protections sought by the industry could—if adopted by Congress—ultimately harm consumers. One would allow broadcasters to retain both channels in cities until 95 percent of the homes used digital TV. That threshold, given that digital TV will be unaffordable to many, is ridiculously high. Broadcasters could double-dip on public airwaves for years.

Another measure could delay the advent of digital television, a needless interference with the market. To be sure, the Federal Government has a responsibility to ensure that the transition from analog to digital is orderly and affordable. But artificially slowing the move could have drastic effects, from smothering competition to forestalling the convergence of telecommunication technologies. Again, the people lose.

The Federal Government needs to speak with a single voice. Congress should join hands with the Clinton administration and hold broadcasters to the 2006 deadline.

Cutting a deal now would weaken the leverage of the Federal Communications Commission, just as the information and entertainment industries awaken to the new digital age. By siding with broadcasters, Congress would lose a fair and painless way to generate billions for the Treasury, and leave behind a golden opportunity to create a long-term trust to fund public broadcasting. Such short-sighted policy serves neither political party. It undermines the future of digital television. It certainly is unfair to taxpayers.

The industry maintains that government would still own the second channels, and that, technically, is true. But extending the free ride would likely drive down the market price for the returned channels. Why should taxpayers take a hit on a booming market—one the government has generously stoked already?

Americans have proved, over generations, their insatiable desire for better TV. Going digital is not—in a business sense, at least—the leap of faith that some broadcasters would have us believe. What's more, fudging the deadline could effectively keep some communities from upgrading their communications for law enforcement. Security blankets are fine. But this one broadcasters cannot justify, and Congress cannot afford.

[From The New York Times, October 11, 2000]

NETWORKS RIDE FREE, DELAY DIGITAL

(By William Safire)

WASHINGTON—What powerful special interest strikes terror in the heart of both parties in Congress, and turns both Al Gore and George Bush into quavering sycophants?

In terms of ripping off the taxpayers with not a peep from the news media, nothing compares with the broadcasters' lobby. This phalanx of freeloaders has stolen the free use of great chunks of the most valuable natural resource of the Information Age: the digital television spectrum owned by the American people.

Five years ago, despite warnings of John McCain, Bob Dole and former Federal Communications Commission Chairman Reed Hundt, NBC, ABC and CBS pulled a bait-and-switch. Because their analog spectrum, a gift to them from the past, was outdated, they demanded a lion's share of the new, digital bandwidth.

When a few of us suggested that this national resource be opened to competitive bidding rather than given away, the broadcasters insisted that the airwaves were their entitlement. With a gift of the new spectrum, they promised to deliver free television broadcasts on high-definition television.

The Republican Congress and Clinton White House promptly doubled the broadcasters' bandwidth—a freebie estimated then at $70 billion, now worth far more.

Worse, the lobby was told it could keep making money on its old analog portion of the spectrum until 2006, or until 85 percent of American homes have digital TV, whichever is later. But it took more than 20 years for color television and 16 years for video recorders to reach that level of market penetration. That's like giving the broadcasters squatting rights on the digital spectrum for decades to come.

Result of Congress' foolish and craven gift of such a cost-free option? Broadcasters have been sitting on their hands, delaying new development and looking for ways to use the new spectrum for profitable cell phones and wireless e-mail, which has nothing to do with broadcasting the promised free digital TV.

Meanwhile, cable and satellite companies, having invested heavily in digital technology, provide the new wares to consumers—but at a high price. U.S. taxpayers, who invested $70 billion of spectrum value in broadcasters to get free digital TV, are forced to wait for decades. Lesson: When private money is on the line, private companies move fast; but when public assets go to private pockets, at no interest, private companies sit tight.

William Kennard, chairman of the FCC, uses a homely analogy about spectrum squatters: It's as if Congress gave each broadcaster two rent-controlled apartments on Manhattan's Upper West Side, and the broadcaster occupied one while leaving the other empty.

What's the FCC to do when Congress and the White House refuse to say "use it or lose it" to the squatters—and thereby let a lobby threaten the U.S. lead in new technologies? To speed our transition to free digital TV, Kennard will mount the bully pulpit in a New York speech today.

He'll call on Congress to require that all new television sets be DTV-capable in 2 years. High volume would not only lower the price of receiver chips to manufacturers but also stimulate consumer demand for the improved images—which, in turn, would provide the profit incentive to broadcasters to get off their duffs.

Then the FCC chairman will urge Congress to close the 85 percent loophole that now turns the double dose of spectrum into a generation-long broadcasters' entitlement to corporate welfare.

Then he'll suggest requiring a fee after 2006 for the use of the old analog channels. "This 'spectrum squatter's fee,' Kennard said, "would escalate yearly, until broadcasters complete their transition to digital and return the analog spectrum to the American people."

That would light a fire under the networks and even encourage debates at public-dispirited NBC.

Although the subsidized industry's legion of lobbyists will lash back in fury, now's the time to ask: How will Al Gore, the professed populist, handle this hot potato?

Where stands George Bush, who would probably appoint the FCC commissioner Mike Powell, Colin's son, to the chairmanship?

Let's find out if either candidate would propose legislation to stop the giveaway and to sell or lease the public's spectrum—thereby bringing free broadcast digital TV to average Americans. Or would both let the huge ripoff roll?

[From Electronic Media, February 3, 1997]

SPECTRUM 'PORK' BLASTED—AUCTION THREAT LOOMS

(By David Hatch)

WASHINGTON—Broadcasters got another chilly blast on Capitol Hill last week when Sen. John McCain, R-Ariz., vowed to eliminate "broadcasting spectrum giveaways" as part of a larger congressional effort to trim "corporate pork" from government spending.

The senator specifically said the government should not give digital television spectrum to broadcasters for free because such "corporate welfare" would cheat taxpayers out of tens of billions of dollars in potential spectrum auction revenue.

The powerful chairman of the Senate Commerce Committee also noted that the White House's fiscal year 1998 budget proposal, to be unveiled next week, is expected to call for spectrum auctions.

But he told reporters that the proposal likely will recommend only that "generic" spectrum auctions be held over the next several years to raise between $25 billion and $35 billion.

The White House isn't "going to be so dumb as to specify" which spectrum should be auctioned, he said, adding that the "heavy lifting" will be left to Congress.

Meanwhile, Mark Buse, an aide to Sen. McCain, estimated last week that auctioning the digital TV spectrum alone could generate between $14 billion and $30 billion "over time." At the press briefing, the senator made clear that he's willing to negotiate with the industry and strive for an accommodation on auctions.

"I realize that the broadcasters have a very powerful lobby here in Washington," he told reporters. "I want to get as much as I can for the taxpayers on this (issue), so I'm willing to sit down and negotiate."

Broadcasters remain opposed to auctions and maintain that digital TV spectrum would not be distributed for free-it would be "loaned" to them because they'd be obligated to return their analog spectrum, which would later be auctioned.

Regarding the targeting of corporate pork, Sen. McCain and Rep. John Kasich, R-Ohio, chairman of the influential House Budget Committee, are heading up a bipartisan coalition of senators and congressman that supports creation of an independent commission to review potentially wasteful government programs.

To that end, Sen. McCain and Sen. Russ Feingold, D-Wis., last week introduced a bill to create such a commission, whose members would be appointed by the president and congressional leadership. The commission would make recommendations about program cuts that could be included in the White House's fiscal year 1999 budget.

Rep. Kasich plans to introduce a similar bill in the House in the near future, a congressional staffer said.

Despite his strong rhetoric, Sen. McCain still has some convincing to do among his colleagues. Rep. Kasich and other House Republicans held a separate press conference last week to unveil a list of pork programs they'll target, but spectrum giveaways weren't on the list.

[From the Grand Rapids Press, Sept. 17, 1997]

THE NETWORKS' FREE RIDE—CONGRESS SHOULD TAKE BACK CHANNELS GIVEN TO TELEVISION BROADCASTERS

Television broadcasters owe American taxpayers $70 billion. That's the value of new TV channels they received without cost earlier this year in exchange for their pledge to develop high-quality digital television. Today, the digital TV promises are emptier than the programming the networks offer.

Gullible Federal lawmakers and regulators should wise up. They must either demand payment for the channels or take them back and auction them off as should have been done at the start. If Congress has no use for the money, taxpayers certainly would.

The issue revolves around high definition television (HDTV). Proponents say that HDTV provides a far superior picture than the existing analog signal. In April, the Federal Communications Commission agreed to give away TV bandwidths to broadcasters after the TV companies promised that all their programming would be in digital form by 2006.

A not-so-funny thing happened on the way to television nirvana. Several broadcasters, led by ABC, have abandoned HDTV. Instead they will chop up the spectrum given to them and create as many as a dozen lower quality digital channels. Adding insult, the new channels probably won't be available to the general public. The networks are likely to scramble the signals so they are viewable only a pay basis, like cable television.

Other broadcasters say they intend to offer digital television, but they wont meet the 2006 target because of the cost. Billions must be spent in new equipment. Local television stations are expected to absorb the largest chunk because they will have to modify transmission towers.

A compromise has been proposed by the Public Broadcasting System. PBS pledges to broadcast two to 3 hours a day of HDTV. But what good is that? A digital television is likely to cost about $3,000 when first introduced. Anyone who would pay anything close to that amount with only 2 hours a day of digital programming ought to be locked up.

Congress must step in and order the FCC to stop giving away digital airwaves. Then it should revoke the channels already given away. Michigan has three members on the House Commerce Committee, which oversees the FCC, who should use their influence: John Dingell, D-Dearborn, Bart Stupak, D-Menominee, and Fred Upton, R-St. Joseph. Broadcasters use the airwaves at the discretion and pleasure of the public. There is nothing pleasing about the way they finagled free channels.

Senator FITZGERALD. Thank you. And I would also like to read from a letter to the editor that Bob Dole wrote to the *New York Times* on Thursday, March 27th, 1997, because I think he summed up the situation pretty well.

He said, "the Clinton Administration, Congress, and the Federal Communications Commission are about to make the already difficult job of balancing the Federal budget harder than it has to be. In just a few days, the FCC is going to give away the first broadcast licenses for digital television to broadcasters for absolutely nothing. If the government sold this new spectrum, it could be worth $12 billion to $70 billion. The network news programs are proud of their commendable watchdog segments like 'The Fleecing of America,' 'Reality Check,' and 'Your Money.' But the networks and many newspapers that own TV stations have largely ignored their own fleecing of the taxpayers."

Now to the gentlemen who are here on the witness stand, wireless phone and direct broadcast satellite companies have paid for airways to upgrade or offer new services. Are you willing to pay anything at all for this spectrum that has been given you, particularly if you are going in 2006 to say, to rely on this exception, that 85 percent of the homes don't have the digital TV, therefore under that exception we should be able to keep our old analog spectrum too. Are you willing to pay anything at all to the taxpayers of this country?

Mr. WILLNER. Senator, as a spokesman for the cable industry and the operator in Rockford, Illinois where we spent $30 million to rebuild that cable system from soup to nuts in order to deliver digital television to our subscribers there, the industry has already paid $42 billion in upgrading its plant and we are 75 percent of the way finished. So, the cable industry has put its money where its mouth is, and we are delivering digital.

Senator FITZGERALD. How about the broadcasters?

Mr. TUCKER. Senator, first of all, that $70 billion giveaway, I can tell you that there are no local broadcasters that have benefited from that and we have not enlarged our bank accounts at all. All we have done to make our date right now, and to provide digital service to the communities, HDTV or SDTV, is spent money to convert our plants.

We are a free over-the-air service, and the only way we raise any money or get any money for our business is by creating advertising revenue.

Senator FITZGERALD. But that spectrum has value, right? The government could have auctioned it off, others would have bought it for cell phone operations or for any variety of communications. Is that not correct?

Mr. TUCKER. Yes, sir, it is. I do not think it is ours permanently, I think it is on loan to us during this transition only.

Senator FITZGERALD. But, do you expect to be loaned permanently for free both the analog and the digital spectrum, or do you anticipate you will be giving up the analog spectrum?

Mr. TUCKER. We anticipate giving it back, sir.

Senator FITZGERALD. You do?

Mr. TUCKER. Yes.

Senator FITZGERALD. But only at such time that 85 percent of the country has digital television sets, which we are finding out is going to be a very long time.

Mr. TUCKER. Senator, I do not think that any of us want to turn off the analog signals and disenfranchise a whole group of Americans. So until we get those sets deployed and in the houses, no.

Senator FITZGERALD. If in 2006 you still have the digital and the analog, what would you think if Congress proposed that you pay a rent for keeping your analog stations?

Mr. TUCKER. All we are doing right now is running dual plants, sir.

Senator STEVENS. I could tell you, Senator, you could not do that unless you canceled their licenses. They have got them now.

Senator FITZGERALD. Well, that is what I am saying we do. They are supposed to give them back by law.

Mr. TUCKER. Senator——

Senator FITZGERALD. You do not want to pay rent?

Mr. TUCKER. No, sir.

The CHAIRMAN. The problem here is going to the Congress to get an extension, and that is what you are planning on doing.

Senator FITZGERALD. Mr. Sagansky.

Mr. SAGANSKY. Senator, our business is free to the consumer. What we get paid for is delivering eyeballs. We sell advertising.

So right now we are $2 billion into probably an $8 billion digital buildout, and we haven't received one penny. We would like nothing more than for this digital buildout to be over, because then we will be able to give back the analogs to the government, but more importantly, for the consumer, it is going to be a much better television experience. They are going to get more content and better pictures.

Everywhere else in the world that it has happened, the consumer has enjoyed palpable benefits. So we want to get this thing over with. We are spending money right now and we are not getting any return whatsoever. That is not in our interests. That is not in anybody's interest.

Senator FITZGERALD. Well, I just think had I been here in 1996, 1997, I would have strongly urged the government to auction off this new spectrum, and I think that would have been a fairer way to do it than to determine by raw political clout who gets to own this new spectrum.

I am hopeful that we will come to some resolution that would allow you to have your digital spectrum but make you pay something if you are keeping the digital spectrum along with the analog spectrum in 2006. I think that is only fair to the taxpayers. But thank you for being here.

The CHAIRMAN. Thank you very much.

We will have the next panel, which is Dr. Mark Cooper, the Director of Research for the Consumer Federation of America; Mr. James Gattuso, Vice President for Policy and Management of the Competitive Enterprise Institute; Dr. Joseph Kraemer, Director, LECG, LLC; and Dr. Tom Hazlett, Resident Scholar, American Enterprise Institute.

Would everybody take their seats please so we can continue with—including the member of the press standing there and the gentleman talking to him.

Dr. Cooper, welcome. Thank you very much.

STATEMENT OF DR. MARK COOPER, DIRECTOR OF RESEARCH, CONSUMER FEDERATION OF AMERICA

Dr. COOPER. Thank you, Mr. Chairman, members of the Committee.

Almost halfway through the transition to digital TV we find that considerably less than half of the programming during considerably less than half of the viewing hours on considerably less than half of the stations in considerably less than half of the markets in this country have a digital product available. Now when you do the math, you will discover that less than 5 percent of the product space has been populated by the broadcast industry.

Clearly, TV time is too slow for the digital age. And there is no chicken and egg problem here as referred to earlier. The first mover risk that the broadcast industry was supposed to take was compensated more than adequately by the giveaway of spectrum. But broadcasters, having received the second or third most valuable real estate in the digital media economy, want more concessions.

They want guaranteed must-carry, low charges for ancillary service fees, and they want consumers who have paid the opportunity costs of giving the spectrum away, to pay real dollars up front to buy tuners for programming that doesn't exist. They want consumers to buy, pay for tuners that will be useless 95 percent of the time.

Now we are confident that if they had not received this asset for free, they would have been much more quick to develop it. In other words, the digital spectrum is a grossly underperforming asset that was mismanaged at the outset by Federal policymakers and is being mismanaged by the corporations that now control it.

In the same 5 years since we began this debate, in which digital TV has populated no more than 5 percent of the product space, cellular phones have moved from 15 million to 100 million customers. The Internet has moved from about 5 million to over 50 million customers. In other words, there is immense demand for channels of communication in the digital age, and the TV industry is moving too slowly.

We need to go in a different direction. Spectrum is a public resource. The right to use a channel of spectrum is a monopoly granted to broadcasters who are allowed to use that space and exclude others from that space, and they haven't paid for it.

Moreover, because television is the dominant means of communications and dissemination of information in our society, it has always borne special pubic interest obligations. As higher quality and interactive TV intensified its reach, immediacy and impact, given the greater and greater power to influence, educate and communicate, the consumers of this country believe that the public interest obligations on this new powerful medium should be expanded.

It is time for the public to get full value for its very very valuable property. Let us relieve the broadcasters of their burden of trying to figure out how to use this space. Let us license it at an auction price to the highest bidder, for the freedom to use it however they want for a limited time sufficient to recover their other investments, and capture that full value, economic value. Let us take the proceeds from those auctions and put them in a trust fund used solely to develop civic non-commercial programming based on public interest and culturally relevant content that is locally developed to fill a very clear need in this country for that kind of responsive programming.

Third, part of the spectrum should be set aside for airing that civic non-commercial programming.

Fourth, development of the spectrum should insure universal availability of digital pictures.

And fifth, maybe it is time to use part of the spectrum to experiment with entirely new ways of exploiting this very valuable public space.

We believe this is the perfect time to expand the public interest use of this new medium. The broadcasting industry is exactly the right place to start, because it is so completely reliant upon a public asset, the airways. We should do this early in the process before it becomes filled up with commercial applications which will make it even more difficult for non-commercial applications to find a space.

The broadcasters have had their chance and it looks like they blew it. They will not meet the deadlines, they have gone on strike, and they will come back and force this Congress to let them out of their obligations. It is time to subject this valuable public resource to a market test. It is time to rent the digital spectrum to the highest bidder, rent not sell, for a timeframe sufficiently long to allow reasonable development and recovery of capital costs, and apply those proceeds directly to the meaningful public purposes that are much more commensurate with this extremely powerful and valuable new means of communications, entertainment and education.

Thank you.

[The prepared statement of Dr. Cooper follows:]

PREPARED STATEMENT OF DR. MARK COOPER, DIRECTOR OF RESEARCH, CONSUMER FEDERATION OF AMERICA

Mr. Chairman and Members of the Committee, today we are evaluating whether the broadcasters have been good stewards of a windfall they received over half a decade ago—free use of the digital television spectrum in exchange for a promise that they would rapidly roll out valuable new services to the public. We conclude that the broadcasters have not been good stewards. While they inch toward rollout of digital television, there are hundreds of innovators—operators with business models that could provide real competition across a wide range of communications and entertainment industries—who are being denied the chance to compete.

The public must be given something in return for the windfall that the broadcasters have received. We could take this opportunity to do one of three things: (1) compensate the public by setting aside some of spectrum for noncommercial use, and applying the proceeds of an auction of the rest of this spectrum band for a public interest communications trust fund, (2) enact mechanisms to ensure that the broadcasters roll out new digital services expediently, (3) open this spectrum to innovative, efficient new competitors who could eliminate monopoly leverage across the full range of communications industries.

However, instead of compensating the public, there are proposals on the table to make the public bear the costs of additional windfalls to the broadcasters—by requiring that new television sets include digital tuners, forcing any consumer wishing to buy a television to pay extra; by requiring cable companies to set aside more space for broadcasters; and by charging broadcasters an extremely low fee for use of spectrum to provide ancillary services. As I hope to make clear, these are out-of-pocket and opportunity costs that the public should not be forced to bear.

When the American broadcasting industry was given use of the digital broadcast spectrum at no charge over half a decade ago, the give-away was controversial for at least two reasons. First, we must remember that the right to use a channel of spectrum is a monopoly given to an individual where that individual and no one else has the right to convey information. In other words, the broadcasters got the right to communicate in these channels and the right to exclude others from communicating in these channels. And they got that at no cost. Of course, they did not "build" the spectrum—the airwaves existed long before the broadcasters. Consumer Federation of America and Consumers Union were dismayed that private corporations would be given exclusive rights to transmit in the digital television spectrum without paying for it, and without adequate financial incentives to rapidly put it to meaningful use.

The second reason this give-away was so controversial was because broadcast television is the dominant means of disseminating information in our democratic society, it has always borne special public interest obligations. As higher quality and interactivity intensify television's broad reach, impact and immediacy, giving it even greater power to influence, educate and communicate, CFA and CU believe that the public interest obligations it bears should expand, or at least traditional obligations should remain. At the time of the digital spectrum give away, no such obligations were imposed.

Unfortunately, when entrepreneurs receive valuable public assets for nothing, they do not have the normal economic incentives of competitive market players to meet consumer needs in an efficient manner. The effort that goes into exploiting assets generally reflects their underlying costs to the firm, not their value to the public. In the case of broadcast spectrum, which had a price tag of zero, the broadcasters have not shown themselves to be worthy stewards of this valuable resource.

While corporations have sat on their asse(t)s, Federal policymakers have also failed to move aggressively to define the nature of the public interest obligation that would be attached to this valuable windfall. Lacking the compulsion of investment at risk or public obligations, the broadcasters have moved very slowly in developing the programming that will fully exploit the value of this resource. We are confident that if these corporations had been forced to pay for spectrum, they would have moved much more quickly to exploit its value.

Five years of wasted opportunity imposes a heavy cost on the public, especially in cybertime. The digital spectrum is a grossly underperforming asset that was mismanaged at the outset by Federal policymakers and is being mismanaged by the corporations that control it. In the 5 years that the digital spectrum has been underutilized, the number of cellular telephone users in this country has increased about eight fold, from around 15 million to around 100 million. The number of households on the Internet has increased more than ten fold, from about 5 million to well over 50 million. In other words, the demand for open channels of communication is increasing rapidly while the broadcasters move at a snail's pace. In a proceeding currently before the Federal Communications Commission, the Commission is considering stripping educational users of their spectrum to accommodate the burgeoning demand for new wireless services. That educational users should be displaced while broadcasters sit on idle spectrum seems to be the reverse of what ought to happen.

Confronted with the industry's failure to take off, the Federal Communications Commission (FCC) has been searching for solutions. One very bad idea it is considering is to make the public, which has already borne the opportunity cost of giving the spectrum away for free, pay for digital sets before there is adequate digital programming. The FCC is considering easing the way for broadcasters by forcing equipment manufacturers to install digital tuners, the cost of which will certainly drive up the price of new television sets.

While there is an inevitable "chicken and egg problem" with any new broadcast technology, the only possible point of the give-away was to compensate the broadcaster for their first mover, "chicken or egg" risk. Having been given the most important input, they were supposed to develop the programming, which would pull consumers to the new product. At the inception of television, the industry did not develop because millions of consumers went and bought expensive television sets before any programming was available. It developed because programming was available and it was something that consumers wanted. But right now the FCC is contemplating making the consumers ante up again, with no guarantees that the broadcasters will live up to their part of the new bargain. We should not tell consumers "if you come, we will build it." Instead, if the broadcasters build what they promised, consumers will come.

It is time to revisit past policy mistakes regarding digital television. The public owns the spectrum and it should get full and immediate value out of it in four ways.

1. Licenses should be auctioned off to the highest bidder with the freedom to use the spectrum for a limited time for the use the highest bidder values most.
2. Proceeds from those auctions should be placed in a trust fund used solely to develop civic, noncommercial programming, based on public interest and culturally relevant content, locally developed.
3. Part of the spectrum should be set aside for the airing of that civic, noncommercial programming, including a set aside for candidates for public office to air their views prior to elections.
4. Development of the spectrum should ensure universal availability of the digital media.

Let me stress that we believe this is exactly the right time to expand the public interest obligations of all the digital media, and broadcasting is exactly the right industry to start with because it so clearly relies on the use of a public asset. As communications, computers and entertainment converge in the digital media economy, we frequently hear the claim by the companies which dominate these component industries that public interest obligations must be abandoned. Each of the industry segments that is converging points to a public interest obligation that it bears, which its competitors do not bear, and claims that it must be excused from that obligation. This race to the bottom obliterates all compensation for public assets and public interest obligations.

We take the opposite view. Each of the industries relies on a public resource in some fashion, spectrum or right of way, and is imbued with the public interest. Convergence should improve performance in all respects, including raising the level of civic and political discourse, not lowering it. We should have a highest common denominator in which the converged media takes on the public interest obligations of each of the component industries and perhaps even adds some to reflect the increased power and impact of the new digital medium. It should come early in the process, before commercial applications fill up the expanded digital product space and place even greater pressure on the educational, civic and culturally uplifting programming that is not as commercially attractive. In short, we support efforts to extract full value from public resources and we think that a part of that value must be realized through payment for use of the digital spectrum and through fulfillment of expanded public interest obligations.

We are certain that the broadcasters will moan and groan about how much it costs them to add other assets necessary to use this spectrum, but we are skeptical about these claims. The most that this could mean, even if it were true, is that policymakers have allocated the spectrum to the wrong uses, because there appear to be many other entities that are more than willing to pay for spectrum to bring digital products to the market.

The broadcasters had their chance and it looks like they blew it. It is time to subject this valuable public resource to a market test. It is time to rent the digital spectrum to the highest bidder (rent, not sell, for a timeframe sufficiently long to allow the reasonable opportunity to recover investment costs) and apply the proceeds directly to meaningful public purposes that are more commensurate with the full value of this new rich, powerful and influential means of communications, entertainment and education.

The Consumer Federation of America is the nation's largest consumer advocacy group, composed of over two hundred and forty State and local affiliates representing consumer, senior-citizen, low-income, labor, farm, public power and cooperative organizations, with more than fifty million individual members.

Consumers Union is a nonprofit membership organization chartered in 1936 under the laws of the State of New York to provide consumers with information, education and counsel about goods, services, health, and personal finance; and to initiate and cooperate with individual and group efforts to maintain and enhance the quality of life for consumers. Consumers Union's income is solely derived from the sale of *Consumer Reports*, its other publications and from noncommercial contributions, grants and fees. In addition to reports on Consumers Union's own product testing, *Consumer Reports,* with approximately 4.5 million paid subscribers, regularly carries articles on health, product safety, marketplace economics and legislative, judicial and regulatory actions which affect consumer welfare. Consumers Union's publications carry no advertising and receive no commercial support.

The CHAIRMAN. Thank you, Dr. Cooper.

Mr. Gattuso, welcome.

STATEMENT OF JAMES L. GATTUSO, VICE PRESIDENT FOR POLICY AND MANAGEMENT, COMPETITIVE ENTERPRISE INSTITUTE

Mr. GATTUSO. Thank you. Advanced television is a technology that has long been in development, long on the public policy plate, and long promised to consumers. Looking through my files this week, I found several studies like this one from the Congressional Budget Office from 1989, predicting widespread use of HDTV by the late 1990s. One part of the study showing 10-, 12 million users by the year 2001. Many times we have been disappointed with this technology.

Despite all these delays, digital television was authorized and allocated by the FCC under congressional guidance in the late 1990s. In a departure from recent practice with other new services, however, no competitive bidding was used to determine the licensees for this new service. Instead, as this Committee well knows, the assignments for the new digital channels were given to existing broadcasters on the basis of a loan, a loan that was supposed to be a short-term situation for this transition to take place.

Today however, the status of that loan is in doubt. Based on current adoption rates, digital television is extremely unlikely to achieve the 85 percent penetration rate by the year 2006 called for by legislation.

This is not to say that DTV has been a total failure. As the Consumer Electronics Association points out, sales of DTV units have increased substantially in the last year or so, with some 600,000 total units sold in the year 2000. Just a couple of days ago CEA announced about 81,000 units sold in the month of January alone. So the units are being sold, but that's not the whole picture. There are several caveats to those numbers.

The most important is that most consumers buying DTV units are buying monitors instead of integrated sets that will allow them to receive broadcast signals without a set-top box, and the number of set-top boxes being sold is very very small. As a result, although a lot of people have DTV units, the number of people watching digital broadcast is still extremely small. It's minuscule.

I think this is what led *Broadcast and Cable* magazine not too long ago to say that digital television has become one of the leading cocktail party conversation killers among broadcasters. It's something that's not working out, we do not have the excitement over it and the optimism over it, that might have seen in prior years.

Even using the more optimistic numbers as to sales, as to total units, we are not going to reach the 85 percent penetration rate. We are far from it. That is basically not going to happen.

Now this delay is of particular concern to taxpayers and consumers because of the potential value of the spectrum in alternative uses. Third generation wireless, Internet access, all sorts of new technologies are on line. In terms of monetary value, it is hard to estimate, but late last year the re-auction of the C-block spectrum alone raised some $17 billion. We are talking about a substantial amount of money here.

So what should the public policy response be? A number of approaches have been already proposed. One that you have heard earlier today would be to increase content regulation of broadcasters. I think that is the wrong approach. First, it doesn't address the basic spectrum problem that we are facing, that 12 MHz spectrum now being used for television is not available for other uses.

Content regulation could also drive viewers away from over-the-air broadcasting to other media that are not so regulated, the opposite of what presumably we want to do. And also, there are non-trivial free speech questions which should not be ignored.

Various forms of economic regulation have also been proposed, including—as has been discussed today—whether new receivers should be required to accept digital signals. As a practical matter, such a requirement would impose significant costs to consumers, as much as several hundred dollars. $200 to $300 is what I have heard.

More broadly, despite all the promise of DTV, there is no guarantee that consumers ultimately will prefer it at all. A decision should not be forced on them by policymakers.

Another option would be to simply require the return by broadcasters of the analog on the original date of December 31, 2006. This would hold broadcasters to the original agreement to return the frequencies and make this spectrum available for other uses. One problem with this is that a mandated end to analog broadcasting will also put the government in the position of picking technological winners and losers. Millions of consumers—arguably having rejected DTV in the marketplace—would be mandated to convert to another technology.

It may be possible to address this problem by simply terminating current analog licenses, but allowing private negotiations to allow analog broadcasting to continue. Under such an approach, if analog broadcasting was sufficiently valued, more so than other wireless services, it could continue. If consumers found alternative wireless services more valuable, then analog broadcasting could be discontinued. It would be a market test.

An alternative marketplace approach would involve providing incentives to broadcasters to vacate spectrum rather than having them pay to remain. Under a voluntary band clearing mechanism recently adopted by the FCC, broadcasters are encouraged to negotiate with potential new wireless licensees on that spectrum to vacate their frequencies.

This voluntary approach seems to create a win-win situation for all involved. The new wireless licensees receive access to spectrum more quickly, allows consumers to more quickly benefit from the services, and broadcasters that do enter into agreements receive payments that could be used to finance their transition into digital services. Consumers gain on both ends. Currently this policy is in effect for channels 60 to 69 and could be extended to other bands as well.

In conclusion, the debate over advanced services has been a long running one throughout the FCC and Congress. At the moment, it does seem likely that there will be an extended transition period, causing valuable spectrum to be misallocated and consumers deprived of potential new services and more valuable new services. The answer to the problem is not new regulation to punish broadcasters or to mandate the use of preferred technology. Instead, policymakers should look for ways to use market mechanisms to insure the best use of spectrum resources.

Thank you.

[The prepared statement of Mr. Gattuso follows:]

PREPARED STATEMENT OF JAMES L. GATTUSO, VICE PRESIDENT FOR POLICY AND MANAGEMENT, COMPETITIVE ENTERPRISE INSTITUTE

Good morning. I am pleased to be here today to discuss Federal policies concerning the transition to digital television. I am Vice President for Policy and, Management at the Competitive Enterprise Institute, a non-profit, 501(c)3 public policy organization that focuses on regulatory policy. Founded in 1984, we have been active on a wide range of technology policy issues. I previously served at the Federal Communications Commission, where I was Deputy Chief of the Office of Plans and Policy from 1990 to 1993.

Advanced television is a technology that has long been in development, and on the public policy plate for many years. Members of the committee may remember the hype that accompanied advanced television during the 1980's, when many touted it as the greatest technological advance since the invention of television itself. It was also seen by many at that time as a necessary step to keep pace with the Japanese, who were perceived as having a significant lead in the technology. Fortunately for the U.S., however, we did not jump into advanced television at that time, as the technology was then analog based—digital advanced television did not become available until several years later. We narrowly missed being locked into an obsolete technology.

Under congressional guidance, the FCC allocated frequencies and set standards for digital advanced television in the late 1990s. Aware of the dangers of being locked into a specific technology, the commission wisely did not mandate that "high-definition television," using the highest level of resolution be used. Instead, broadcasters would be allowed to provide other services to their viewers as appropriate, including the possibility of multi-casting multiple channels of programming, or providing simultaneous data, transmissions.

In a departure from recent practice with other new services, however, no competitive bidding was used to determine the licensees for this new service. Instead, licenses were assigned to existing broadcasters. Broadcasters would then hold two, licenses—their existing "analog" license and a new "digital" license for a transition period. Nominally, this transition period was limited—to expire on December 31, 2006, but an extension was required by statute if 85 percent of households in a market did not have access to digital television by that date, either directly or through a multi-channel provider, such as cable.

As you know, there was considerable opposition to this plan.[1] On equity grounds, this plan represented a transfer of a immensely valuable resource, worth tens of billions of dollars, free of charge to the broadcast industry. Perhaps of even greater concern were the economic concerns that by protecting the frequencies from marketplace pressures, it would be less likely to be used as efficiently as possible.

Nevertheless, the plan was adopted, largely on the premise that the additional spectrum provided to the broadcasters was only for a short period of time. It was to be a loan, not a giveaway.

Today, however, the status of that loan is in doubt. Based on current adoption rates, digital television is extremely unlikely to achieve the 85 percent goal by 2006.

This is not to say that DTV has been a total failure. The record has been mixed. In the first year or so that DTV units were available, sales were minuscule. Last year, however, sales increased substantially, with some 600,000 total units sold, according to, the Consumer Electronics Association (CEA). Earlier this week, the CEA announced that January factory-to-dealer sales of DTV units totaled 81,629, a 234 percent increase over last year. It predicted 1.1 million units to be sold in 2001, and 10.5 million to be sold by 2006. This is pretty much what the Consumer Electronics Manufacturers' Association (CEMA) predicted when sales began in 1998, which originally predicted 10 million in sales between 1999 and 2003.[2]

There are some important caveats to these numbers, however. First, the CEA numbers refer to sales to dealers. The number of units sold to consumers is much lower, about 200,000 last year according to one report.[3] In addition, many consumers are buying DTV monitors separately, instead of integrated sets that allow them to receive broadcast signals without a set-top box. The number of such integrated sets sold has only a small fraction of total unit sales. As a result, despite the impressive total number of units sold, the number of people watching digital broadcasts is still extremely small.

Even CEA's more optimistic numbers, however, raise a concern about the digital transition. At that rate, market penetration would almost certainly be far below the 85 percent needed to trigger a return of the analog spectrum. In fact, if the numbers track, CEMA's original projection, consumer penetration would only be at 30 percent in 2006. As a result, we may face a long wait—perhaps decades—before the spectrum "loaned" to broadcasters is returned.

This delay is of particular concern to taxpayers and consumers because of the potential value of this spectrum in alternative uses. The frequencies involved are (in, spectrum terms) prime real estate, and could be employed for a variety of wireless services, including third-generation mobile services. Given the wide variation in auction revenues over the years, putting a specific value on these frequencies is a tricky business, but it is sure to be significant. Last year's re-auction of the "C-block" PCS spectrum, alone garnered some $17 billion.

A number of approaches have been proposed for dealing with this situation, many of them bad. One approach is to increase content regulation of broadcasters. Such, regulation could decrease the value broadcasters receive from the spectrum, in effect, decreasing the size of the giveaway. The problem is that it would also punish consumers, by limiting broadcasters' ability to provide them with what they want. It also raises significant free speech concerns. Government intrusion into content is simply not an answer to spectrum management problems.

Various forms of economic regulation have also been proposed in order to drive consumers to DTV. The FCC, for instance, recently began an inquiry into whether all new receivers should be required to accept digital signals. While such a step was taken in regard to UHF signals, policymakers should always be careful about imposing such mandates. As a practical matter, such a requirement could impose signifi-

[1] Along with others, I testified before this Committee in favor of competitive bidding in March 1996.

[2] Timothy Somheil, "TV or DTV?", *Appliance*, December 1, 1998.

[3] Cited in *Electronic Engineering Times*, December 22, 2000.

cant costs on consumers—as much as several hundred dollars. More broadly, despite all the promise, of DTV, there is no guarantee that consumers will ultimately prefer it. A decision should not be forced on them by policymakers.

Another option would be to simply require the return by broadcasters of analog television licenses on the original date of December 31, 2006. That would certainly be a fair option, for it would merely hold broadcasters to the original agreement to return the frequencies. It would also serve the important goal of making this spectrum available for other uses.

A mandated end to analog broadcasting, however, would also put the government in the position of picking technological winners and losers for consumers. Millions of consumers, having arguably rejected DTV in the marketplace, would be mandated to convert to another technology.

It may be possible, however, to terminate current analog licenses, while allowing the ultimate choice of technology to be left to the market. Broadcasters, for instance, could be allowed to negotiate with the new license holders to continue to use their frequencies for analog broadcasting. Under such an approach, if analog broadcasting were sufficiently valued—more so than other wireless services—then it could continue. If consumers found alternative wireless services more valuable, then analog broadcasting could be discontinued.

An alternative marketplace approach would involve providing incentives for broadcasters to vacate spectrum, rather than having them pay to remain. Under a voluntary band clearing mechanism adopted by the FCC, broadcasters are encouraged to negotiate with potential new wireless licensees on that spectrum to vacate their frequencies. Specifically, the Commission established a rebuttable presumption that such agreements to relocate are in the public interest.

This voluntary approach seems to create a win-win situation for all involved. The new wireless licensees receive access to spectrum much more quickly, allowing consumers to more quickly benefit from those services. Broadcasters are not required to relocate, but will gain the incentive to do so. This incentive would be proportionate to the value of their stations—meaning the least-watched stations would (all things being, equal), the first to relocate, and the most-watched stations the last. And broadcasters who do enter into agreements receive payments that could be used to finance their transition to digital television.[4]

Currently, this policy is in effect for channels 60-69 (and for three-way deals involving broadcasters on other channels). Based on the success of this policy, the Commission will determine whether to extend voluntary band-clearing down the dial to channels 52-59. It is too early to assess the success of this policy, but it looks promising.

Conclusion

The debate over advanced television has been a long-running one for the FCC and for Congress. The issues are complex ones; I know there are no simple answers. At the moment, however, it seems very likely that an extended digital television transition period will cause valuable spectrum to be misallocated, and deprive consumers of valuable wireless services they want and need. The answer to this problem, however, is not new regulation to punish broadcasters or to mandate use of preferred technologies. Instead, policymakers should look for ways to use market mechanisms to ensure the best use of spectrum resources.

The CHAIRMAN. Thank you.

Dr. Kraemer.

STATEMENT OF DR. JOSEPH S. KRAEMER, DIRECTOR, LECG, LLC

Dr. KRAEMER. Thank you. I am Joe Kraemer, an equity partner in a consulting firm, LECG. I am actually responding to questions that were posed by the staff. I was contacted by the Majority staff and the Minority staff and we had consensus, they asked the same question. The question was, given the state of the transition of digital television, can and should government intervene? If yes, how

---

[4] To facilitate such negotiated relocating, one firm, Spectrum Exchange, has already outlined plans to hold a "secondary auction" simultaneously with the FCC's auction of these frequencies. This auction will help bidders ensure that the spectrum they receive licenses for coincides with the broadcasters with whom they enter into band clearing agreements.

can we make an effective intervention? Those are the questions to which I am responding.

If you look at my testimony, I have run three scenarios with respect to digital television transition, taking into account the various factors such as programming, availability, must-carry, and all the issues of spectrum auctions and the like. When you do that, you really have three scenarios.

You have a transition that is fast, and when you look at the rapid transition, the best you do is turn off analog no later than December 31, 2010. You have another scenario which is essentially moderate, that gets you there at December 31, 2015. Then you have a slow one, which is plausible based on just the conversations and the questions we have had today, which gets you out to 2020, probably at which point it is Liberia, Paraguay and the United States still trying to do the transition.

And the real issue is that the actions taken by the government and all the parties, cable, consumer electronics, the programmers—you did not have the networks here. There is a lot of talk about programming, but the networks buy the programming and they were not here. Where is NBC, CBS? You need to bring them in. That is the source. The broadcasters, like NAB, are local licensees. They do not do the network programming.

But you need to take all those into account, and the decisions that you make now really have lead times of 3, 4, 5 years, so if you do not make decisions, you are pushing yourself out to 2015 or 2020.

Now, the question becomes who wins if we accelerate the transition to digital. Well, first of all, it would return the analog spectrum, in which case you have auction revenues. You also have a buildout of the wireless spectrum, which will pump a multiplier effect through the economy which is important, because the telecomm industry in a wired sense is no longer buying technology and you can see the air gap in the economy. We've got to keep the buildout going.

You also shift to a sustaining demand pull. Right now, this is industrial policy which is extremely unusual for the United States. DTV started with government and is being pushed by government. We don't have the DVD, the cellular phone kind of pull out there with the consumers. We've got to move there.

You also can trigger waves of investment by programmers, manufacturers, broadcasters themselves. All these parties are just going sideways and when you do that, you do not have multiplier effects in terms of employment, in terms of the economy. There is a benefit to consumers. Every set of focus groups shows that consumers enjoy and appreciate the improvement in audio and the sort of quality associated with digital. This is something that if they can get exposure to, you could very well trigger a demand for the change to digital and no more hearings will be required.

And also, you should decrease the length of time that broadcasters are operating both analog and digital operations. Broadcasters really do a win-win. Actually, when you run parallel digital-analog plants, it costs you a lot of money and, if you do that for 20 years, you're going to affect your profitability. It isn't logical at an operating level to do both analog and digital for years.

So one question becomes, can government intervene? Yes. Now, if so, can you be effective? And the real question is, can you tip the market? In other words, can you create sustained demand by consumers so we don't need hearings—we really can look on this as a success, not a failure.

In my testimony at page 4, I did have a chart that basically shows what the tipping point would be. What you want to is move that to the left. Can government move it to the left? Can government accelerate the demand for consumers? If you let it drift to the right, then you are talking about 2020 before you turn analog off.

So just looking at it, there are probably two areas where government can intervene—and this is more likely actually the FCC, not Congress. All-channel receivers, there is a logic for mandating those. The manufacturers themselves will not take the initiative one by one, because essentially it puts them in a less-than-optimal competitive position.

If you do that, you will basically move to reduce the embedded base of analog sets. We have 280 million sets in the U.S. that are analog only. We need to decrease that base. We buy 25 million sets a year minimum. You basically mandate there that they have to be able to receive digital; in 4 years you will have probably on average one digital set minimum per household, so you move forward to an early cutoff for analog.

Digital must-carry is also required. There is a lot of infighting, chewing on each other over that. I would suggest you do need to move toward digital must-carry, but sunset it after 3 years. In other words, they carry it for 3 years and at the end of that time the programming should compel the consumers to want it. If the consumers want broadcast digital programming, then the cable folks will put it up and leave it.

And so those are two areas. Obviously there are other issues you need to look at. You need to look at things like programming, but, given that staff only gave me 2 days, this is the best I could do.

Responsibility probably goes to the FCC, not to Congress, just because of the nature of the institutions. So you really have to look at the FCC and maybe you can, you know, talk to Mr. Powell about what he is doing. In the end, should the FCC be dormant on this, you are going to go to 2015 or later. The FCC has got to take the lead, they have got to move out, they have got to make some things happen. Thank you.

[The prepared statement of Dr. Kraemer follows:]

PREPARED STATEMENT OF DR. JOSEPH S. KRAEMER, DIRECTOR, LECG, LLC

Summary of Testimony: Digital Television Transition

I. A range of outcomes is possible. In this testimony and in an attachment hereto, I have outlined three scenarios:

Scenario	Analog Turn-off	Government Role
Rapid	2010	Intervenes Early
Moderate	2015	Largely Passive
Slow	2020	Uninvolved

Actions taken or not taken by Government in 2001 will affect decisively which scenario is realized.

II. It is in the interest of most stakeholders to accelerate the DTV transition (i.e., achieve the rapid scenario). Benefits include: 1. Return analog spectrum leading to auction revenues for the Government and the build out of wireless high speed data networks; 2. Shift to a self-sustaining demand pull market; 3. Trigger waves of capital investment by manufacturers, programmers, broadcasters, and networks which will have multiplier effects on employment and income at each stage in the industry's supply chain; 4. Improve the quality of the TV picture and audio experience for consumers; and 5. Decrease the length of time broadcasters operate expensive dual analog and digital transmission systems.

III. Government intervention is both possible and necessary to accelerate the transition. The critical factor is to expose consumers to digital TV. Once exposed, consumer demand will "tip" the market, creating a self-sustaining mass market.

It is time to consider whether Government can intervene positively and then step aside and let market forces work. Two points of leverage exist:

1. *All channel receivers*: logic exists to enable sets, sold after a date certain to receive over-the-air digital broadcasts; and 2. *Digital must-carry*: with the primary set in 65 percent of U.S. households hooked up to cable, a time-limited (three-year?) requirement to carry both analog and digital over-the-air broadcasts allows consumers to experience DTV.

The lead on intervene probably belongs to the FCC, but Congress has a role as well.

The Critical Issue is the Point When the Market "Tips" and Becomes Self-sustaining

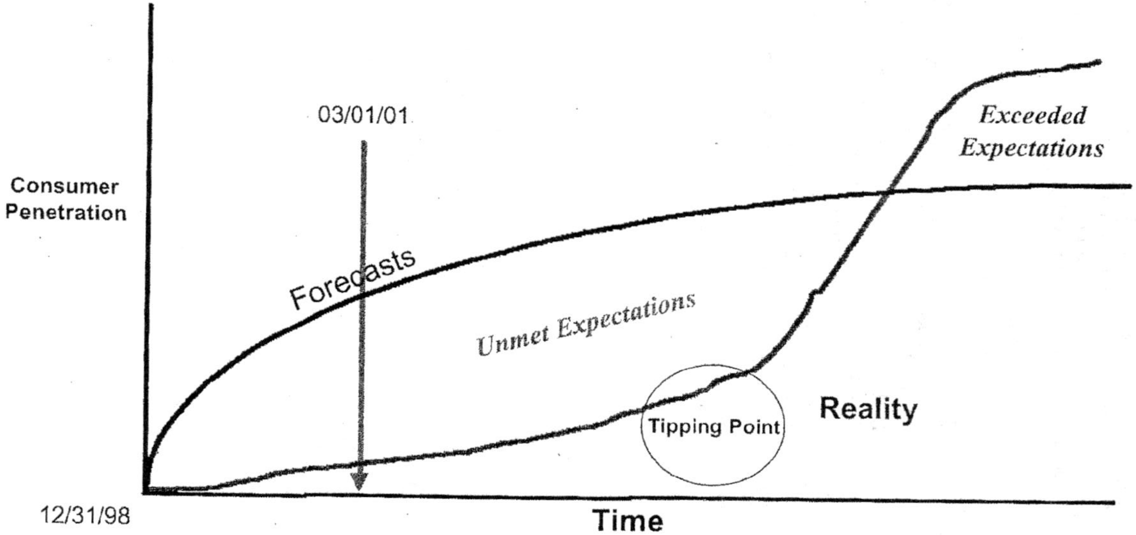

"… We always overestimate the change that will occur in the next two years and underestimate the change that will occur in the next ten." Bill Gates

I. Digital Television (DTV) Transition Scenarios

A. DTV Scenarios

DTV requires a very long-term perspective. The transition to digital could take all or most of the next two decades and will affect literally all 100 million U.S. households.

With respect to DTV, the decisions made in the 2001-02 timeframe have a "long fuse," and a "big bang," three to 5 years later, with a material impact on shareholders, employees, partners, suppliers, customers, and management.[1] Many participants in the DTV transition are playing a game of "bet the company." At a minimum, most stakeholders are placing a significant portion of future earnings at risk.

Scenarios assist decisionmaking under conditions of uncertainty. Scenarios are not intended to predict the future. Rather, they can be used to facilitate an understanding of a reasonable range of options and the consequences of those options. The development of the scenarios used in this testimony are based on the results of interviews conducted in late 2000, as well as a general understanding of industry developments.

In order to be successful, scenarios must be reality based, taking into account external conditions that are "givens" and cannot be changed in the short or intermediate future. For DTV scenarios, it is important to remember that:

1. *No dominant player exists.* The television supply chain is fragmented at each level from manufacturing of equipment through production and distribution of content. No equivalent of Microsoft in the PC operating systems business or Intel in the chip business—or even a duoploy like Coke and Pepsi—exists. Therefore, no single company by itself—not Sony, not General Electric, not Disney/ABC—can determine the outcome. Thus, each stakeholder must formulate their own unique strategies because there is no leader to fall in line behind.

2. *Government is relevant and can affect the speed and course of DTV rollout.* DTV has a political dimension. The FCC, Congress, the courts, and multiple presidents yet-to-be-elected will influence the pace of DTV rollout.

It must be remembered that achievement of the legislated objective of 85 percent of households with digital capability (defined as the primary viewing set) could be attained by some mix of: (1) digital-to-analog cable set-top boxes in combination with digital set-tops for digital sets; (2) satellite digital-to-analog conversion; and (3) free-to-air broadcasts to digital sets with or without a roof antenna. Also, the 85 percent is of primary sets only; it does not address the embedded base of 150+ million secondary sets (that are in addition to the 100 million primary sets in the U.S.).

For DTV three general scenarios make sense:

1. *Rapid Transition:* This scenario incorporates a series of assumptions so that the transition resembles the rapid take up of black and white TV after World War II or the rise in usage of the World Wide Web (i.e., fast, deep, and successful).

Rapid Transition: 85 percent in 2006-08; Analog turn-off 2010-11.

- Stakeholders cut deal to move DTV forward.
- Consumer exposed to DTV and demand "tips" 2005-06 so that mass market emerges.
- Channels 60-69 and 52-59 are auctioned almost on schedule.
- Congress and the FCC intervene on matters such as all channel receivers and must-carry.

2. *Moderate Transition:* The core theme is that the interlocking series of events necessary for DTV go neither terribly right nor terribly wrong.

Moderate Transition: 85 percent in 2010-12; Analog turn-off 2014-15—No stakeholder deal is negotiated.

- Auctions are delayed and not meet expectations; spectrum use taxes are probable.
- Broadcasters operate expensive parallel system both analog & digital.
- Government remains passive and hesitant to intervene.

3. *Slow Transition:* Under this scenario, many factors combine to frustrate and slow the DTV rollout. This could occur due to some combination of technology, regulatory, and/or market factors. (Exogenous events, such as a stock market collapse combined with rising unemployment and declining consumer confidence could also play a causal role.)

Slow Transition: 85 percent after 2014; Analog turn-off 2020.

---

[1] "Long fuse, big bang" decisions involve judgments made, instructions given, and actions taken, the success of which cannot be measured for years but the outcome of which will determine the survival of the organization. One of the ironies of "long fuse, big bang" decisions is that the management that makes these decisions has often moved on and a new generation of managers (and shareholders) have to live with the outcome.

- Free-to-air broadcast TV becomes increasingly less relevant.
- Networks bypass affiliates and go to cable head ends.
- Government takes no action; FCC adopts "let the market decide" attitude.

DTV scenarios do not predict the future. However, they serve to: (1) sensitize stakeholders (including Government officials) to the implications of actions taken or not taken; and (2) emphasize the extent to which stakeholders must cooperate because no single company can control the outcome.

B. Diverse Points-of-View, But Some Consensus on the DTV Transition

As part of an analysis I conducted late last year, broadcasters, manufacturers, network representatives, public officials and industry observers provided facts, opinions, official on-the-record positions, and unofficial not-for-attribution perspectives. Summaries of relevant, key themes that emerged are presented below.

1. *Resolution of certain issues is required to accelerate the rollout of DTV.* The two issues mentioned most often as the most critical to broadcast DTV rollout were: (a) cable must carry; and (b) the availability of high definition and enhanced programming. The logic of the respondents was that, if consumers could see DTV, then this would create demand pull and initiate a market-led transition to DTV. Other factors such as content availability, copy protection, receiver prices, and all-channel tuner requirements must also come into line, or rollout will be delayed.

2. *The core drivers are primarily business and public policy, not technical.* Almost all DTV technical issues have been resolved. Therefore, the issues remaining tend to be: (a) economic—who spends how much and for what return; and/or (b) public policy—should and how can government influence the transition to DTV?

3. *The free-to-air television business will change significantly over the next 5 years.* Over the longer term, 90 percent of primary sets will be wired (either cable or satellite). Therefore, the long-term U.S. free-to-air market will consist primarily of secondary sets (e.g., smaller, largely portable, potentially pedestrian or better speeds), as well as computers (fixed or portable) as receivers.

4. *No single stakeholder controls the rollout of DTV.* A multitude of DTV stakeholders (e.g., consumer electronics firms, networks, local broadcasters, program producers, cable, the FCC) exist with their own business or public policy interests. The potential exists for paralysis through mutually neutralizing business and public policy actions. On the other hand, most of the stakeholders have a shared economic interest in moving the transition forward.

5. *The digitalization of television in the U.S. will proceed; the issue is when, not if.* The rollout of digital video could occur without much of a fixed free-to-air component. Digital production, DVD, satellite, digital cable, and streaming video are accelerating. Local broadcasters remain influential but by themselves are not decisive and could be isolated over the long term, especially if the broadcasters lack consensus on key DTV issues while other stakeholders press ahead with non-free-to-air digital television.

II. Acceleration of the DTV Transition

A. The Benefits of Acceleration

When the history is written, there is a high probability that digital television (DTV) will be compared in some ways to the Internet—slow to take off, dominated in the early market phase by visionaries, benefiting from occasional government intervention, and global in impact but with distinctly American nuances. DTV will also be recognized in retrospect as one of those paradigm shifts that rearrange the economics of entire industries and create lists of winners and losers. Adaptability, flexibility, and a talent for strategic thinking (or lack there of) constitute the three attributes that will separate the former from the latter.

A rapid transition to DTV will: 1. Facilitate the return of analog spectrum that in turn will be auctioned to network operators, which, in turn, will trigger a wave of investment in wireless broadband infrastructure, as well as contribute to maintaining budget surpluses; 2. Decrease the length of time broadcasters will need to operate dual analog-digital transmission systems; no trivial issue for stations in small markets and/or small stations in any market; 3. Shift the basis for the DTV market in the U.S. from the current 'supply push' model (i.e., government compels and broadcasters acquiesce) to a 'demand pull' model that sustains itself as a mass consumer market; 4. Materially improve the quality of the TV picture and audio experience for consumers; 5. Transform the entire TV supply chain from program planning and production through local transmission and reception; 6. Provide a potential new lease on life for the broadcasting industry that has been hemorrhaging viewers for 10 years; 7. Trigger waves of investment spending by manufacturers, programmers, local broadcasters, and TV networks (including free-to-air, cable, and satellite), which will roll through the industry's supply chain with a multiplier effect on employment and income.

If DTV had no other effects other than those above, it would be worth accelerating the transition. However, beyond its first tier effects, DTV will also act as a catalyst and cause second tier effects that will be at least as significant, if not more so. In this second tier, the impacts of DTV will include:

1. Merge the TV and the PC so that the TV will have more in common with today's PCs than contemporary TVs; 2. Double the number of U.S. households with web access to collect information, send/receive e-mail, and shop at home thereby providing television a role in the networked economy of the 21st Century for TV networks; and 3. Intensify competition between and among video suppliers as networks are upgraded for digital transmission which will also provide bandwidth for Internet and other services.

Only a realistic assessment of the situation will achieve the potential of digital television in a reasonable period of time. That promise, by the way, can be more than even the optimists predict, but only if the digital transition is realistically planned and implemented by networks, manufacturers, government, broadcasters, and consumers themselves.

DTV is being rolled out currently without material consumer demand. Consumers that have invested in DTV sets tend to be either: (1) "technophiles" (responding to the potential of digital to merge the TV and the PC); or (2) "videophiles" (emphasizing the improved picture and audio capabilities of digital). These categories constitute the early market. The critical issue is when the DTV market "tips" and becomes a mass market. After the market tips, then it will become self-sustaining and based on 'demand pull' as did other markets such as color televisions, PCs, and cellular telephones.

In order to tip the market (i.e., accelerate the point in time when demand ramps up as a mass market), consumers must be exposed to DTV. Exposure will trigger demand for DTV receivers, digital programming, and ancillary services such as broadcasting to PCs (i.e., shift the market from its current 'supply push' context to a sustainable 'demand pull' basis).

B. The Role of Government

Government and the DTV transition have been inseparable from the beginning. If anything, there has probably been enough government intervention that DTV constitutes a rare example of industrial policy in the United States. The FCC guided the process that developed the DTV standards and then followed congressional guidance when awarding the spectrum necessary to transmit digital programming. At various points along the way (especially on the matter of spectrum award), Congress and the incumbent administration got involved and endorsed or modified private sector and/or FCC decisions as part of the public policy process in the late 1980's and 1990's. Now with the DTV transition slowed, it is time to consider whether government can intervene one last time and, in a positive way to accelerate the transition.

I assume there is: (a) a public policy interest in facilitating a rapid transition to digital television to permit spectrum clearing; and (b) a belief on the part of regulators that market forces should be the ultimate driver of both the growth of digital television programming by broadcasters and acquisition of receivers by consumers. Therefore, if government is to accelerate the transition, then government must accelerate the rate at which consumers are exposed to DTV then step aside and let market forces work. There are two leverage points available.

Both the all-channel and digital must carry requirements would appear to be necessary to catalyze a market-driven DTV transition. If most TV sets cannot receive a digital signal, then there is very little incentive to generate digital programming. Such programming would be almost a novelty as was the case with color broadcasting when there were very few color television sets. However, since about 65 percent of total U.S. homes have cable service, simply equipping the TV sets with the capability to receive digital signals may not provide the necessary incentive unless the cable systems also must carry digital as well as analog off-the-air signals.[2]

Although the FCC is considering the all-channel and digital must carry issues in separate proceedings, the two requirements are interrelated. The first step would be to require that all new TV sets sold be capable of receiving a digital signal. Then, at a date on, or shortly after, the date when all new sets sold must be digital-capable, all cable TV systems would be required to carry both the digital and analog signals generated by the off-the-air stations. This requirement that cable TV systems

[2] For the households with the most sought after demographics by broadcasters and advertisers, cable penetration probably is higher than 65 percent.

carry both signals need only be in place for 3 years or so. After that, market forces would protect the public interest.

1. *Requiring TV Sets to Be Able to Receive Both Analog and Digital Signals.* The causal connection between needing a substantial installed base of TV sets capable of receiving a digital signal before the broadcasters will offer most, if not all, programming in a digital format seems obvious. The need to have such an installed base can be demonstrated by examining what happened to the viewership of UHF stations and the number of UHF stations after the all channel (VHF and UHF) tuner was required for all TV sets. The relative viewership of UHF stations increased among the off-the-air signals. Also, the ease of access of UHF channels and the increased viewership also led to more UHF stations being on-the-air. Finally, this also assisted the emergence of the new networks (e.g., FOX, WB, and UPN). Prior to the All Channel Tuner Act, the UHF stations had a relatively high failure rate and that entry by UHF stations had been very disappointing.

Second, the experience with color television also can be helpful. Color television was never mandated, but color programming was very limited until the installed base of color television sets reached a critical mass. Similarly, one cannot expect a substantial increase in digital programming until there is a substantial installed base of TV sets that can receive digital signals. NBC was seen as taking a substantial risk when it took the lead in going to all-color prime-time broadcasting. At that time, of course, RCA/NBC was vertically integrated into TV set production. Today, even if a network were to make a DTV programming commitment (as CBS appears to be doing), the transition would be stillborn if set manufacturers did not provide follow through with set production.

The argument that making all TVs so that they could receive and process digital and analog signals would raise the costs of these sets substantially would not be true in the case where all TV sets had to have the capability. The engineering and design costs needed to make such a conversion would not be high on a per-TV-set basis if all TV sets had to have this capability.

Nevertheless, without an all-channel requirement, given the highly price sensitive competitive nature of selling the high-volume TV set models, it is less likely that any manufacturer of such sets would take the risk of adding digital reception capability to mass market sets even if the resulting cost per set were low. Even a minimally higher price could be seen as placing the manufacturer at a competitive disadvantage in the mass market. If such a capability were offered only on upper-end sets, the per-TV set cost of offering the capability just on this small subset would be quite high making it unlikely that the price-sensitive customer would purchase such sets.

However, if DTV reception had to be available in all sets, the manufacturers' efforts would be focused on making this capability as low-cost as possible. Further, the costs would be spread over a very large number of manufactured units making the average cost small. New TV sales each year amount to about 25 percent of TV households.[3] Approximately 25 million sets sold annually into an embedded base of approximately 100 million U.S. households. If it were mandated, the manufacturers' efforts would shift to making the capability as inexpensive as possible. There are numerous examples of how offering a feature on all models dramatically reduces the cost of such features and, when a feature becomes standard, that the manufacturers quickly move to reduce costs.

Finally, the high annual sales rate relative to the installed base of TV sets (about 25 percent of households per year) suggests that a large percentage of TV homes would be likely to have at least one digital-capable TV set within 4 years. This would provide a very strong incentive for networks to provide digital programming.

2. *Requiring Cable Systems to Carry Both Analog and Digital Off-the-Air Signals.* The requirement that all cable TV systems must carry both digital and analog off-the-air signals should be implemented no sooner than the date when all new TV sets sold must be able to receive both a digital and analog signal. The implementation might be delayed somewhat because there will be only a small number of TV sets in the base for the first 6 months to a year after the requirement that all new TV sets sold must be digital-capable. It is important to require cable systems to carry both the analog and digital off-the-air signals for at least 3 years. After that, market forces should be relied upon.

At the outset, market forces are not likely to be sufficient. These cable systems would be under some competitive pressure from off-the-air digital signals and possibly from satellite providers (e.g., Direct TV) to carry the digital signals, but these

---

[3] Approximately 25 million sets sold annually into an embedded base of approximately 100 million U.S. households.

providers also may not provide digital "local into local" broadcast programming, limiting the cable operators' competitive incentives to do so.

Again, however, *the market pull for digital carriage needs an initial regulatory catalyst.* If the cable systems do not carry both digital and analog off-the-air signals, then any digital programming generated by the off-the-air stations will not reach the TV sets in cable homes.[4] Given that 65 percent of all U.S. homes are cable TV homes, it would appear essential that cable systems carry the digital signals generated by the off-the-air station to make digital broadcasting valuable for broadcasters.

Cable systems probably will claim that it is not feasible to carry both the analog and digital signals due to channel availability limitations and/or that adding the digital signals is prohibitively expensive. However, digital compression will allow multiple DTV channels to carried within a 6 MHz cable channel. It may be necessary for cable systems to use a converter box to allow the digital signal to be delivered in a form that the digital-ready TV set can process. Such boxes should be ready by the time the must-carry requirement kicks in, or such capabilities could be installed in sets meeting the FCC's "digital cable-ready" specification.

The FCC has asked whether the dual-carriage burden could be reduced by making the dual carriage limitation of limited duration. I believe it would be necessary to mandate only that cable systems carry both analog and digital signals for 3 years after the date when new TV sets sold are to be capable of receiving both a digital and analog signal. At the end of this period, the majority of primary TV sets hooked into cable systems should be digital-capable. Given this situation, market forces would keep the cable system from removing the superior digital signal.

[4] Often, there are TV sets in cable homes that are not hooked into the cable (i.e., get an off-the-air signal). However, the prime-time viewing is most often done in front of the TV sets hooked into the cable system.

BROADCAST DTV ROLLOUT SCENARIOS

	Rapid Transition	Moderate Transition	Slow Transition
Legislation & Regulation *	1. The FCC Chairman adopts DTV as a critical issue for the FCC. 12. Proactive FCC mandates all channel receivers as of date certain (e.g., Jan 1, 2004) for sets 13" and larger. 3. FCC resolves all set-top box technical issues, including copy protection. 4. FCC reaffirms the 2002 free-to-air DTV rollout requirement for commercial broadcasters but allows small markets (e.g., 101 and above) to opt to defer until no later than June 30, 2004.	1. The FCC remains a non-player until the next administration at which time the year 2005 Chairman of the FCC adopts DTV as one of his/her make or break issues. 2. FCC proceeds to rule on/close out open issues as per the rapid transition scenario only 4 years later and with free-to-air somewhat less significant. 3. On-air digital dates for broadcasters stretch out with waivers easy to obtain.	1. DTV not adopted by any administration or FCC Chairman as an issue upon which to spend political capital. 2. Congress holds occasional hearings but becomes irrelevant to DTV. 3. FCC adopts "let the market decide" approach on all key issues. 4. FCC remains reactive not proactive.
DTV Must-Carry	1. FCC resolves cable must-carry (e.g., cable must-carry free-to-air DTV signals up to capacity limits with station election of signal to be carried); program-related enhancements (including advertising and program interactivity) must be passed; HDTV signals must be passed without material alteration; reasonable fees imposed for retransmission of multiplexed programs for which broadcasters charge a subscription fee.	1. FCC delays initiation of must-carry resolution until 2005 (new administration); outcome similar to rapid transition but a half decade later.	1. No mandated free-to-air DTV carriage until analog shut off. 2. DTV carriage prior to analog shut off only pursuant to voluntary agreements.
Consumer Electronics & Set-Top Technology.	1. CE industry reaffirms commitment to U.S. free-to-air DTV; R&D funds committed to improve digital reception; fourth generation chips in sets as of mid-2002. 2. Set prices decline as volume increases consistent with prior CE industry practice. 3. CE industry supports All- Channel Receiver Act as one price of moving DTV forward in the U.S. 4. Cable set-top boxes available with DTV pass through capabilities. 5. Low-cost digital-to-analog converters available at retail stores in late 2004 for unwired sets. 6. DBS and broadcasters deploy joint antenna systems for free-to-air pick-up of digital signal.	1. CE industry puts free-to-air DTV on hold until market more promising; R&D diverted to satellite and cable. 2. Volume ramp up for mass market delayed; probably begins no sooner than 2006. 3. Same as rapid transition scenario except four to 6 years later.	1. CE industry assigns low priority to free-to-air DTV; focuses on cable and satellite markets; R&D funds diverted away from free-to-air improvements. 2. Because of low volume sales, prices decline slowly as sales of free-to-air receivers are minor compared to cable and satellite digital receivers. 3. CE industry gradually and voluntarily installs all channel receivers so that analog-only new sales no longer occur after 2010. 4. Cable operators never make available converter boxes for DTV pass through until analog turn off. 5. Limited retail availability of digital-to-analog converters.

Programming/Content	1. Networks make available significant HDTV programming, particularly sports and movies, as well as other enhanced programming. 2. Local broadcasters use multiplex capabilities to transmit local content (e.g., news and high school sports with channel choice by county). 3. Consumers increase demand for DTV; market pull begins to replace supply push circa 2004-05. 4. Broadcasters sell advertisers on DTV's enhanced capabilities (e.g., interactive advertising and very attractive demographics); advertising revenues increase.	1. Networks delay content; delay rolls through production supply chain delaying digitalization of content. 2. Local broadcasters stretch out multi-casting trials because: (a) lack of must-carry rules frustrate business case; and (b) free-to-air DTV receivers remain scarce. 3. Advertisers focus on cable networks with return channel for interactive advertising. 4. Over-the-air content not in significant in quantity until 2003-04 season.	1. Networks hold production costs down; limited availability of HDTV and enhanced programming until after 2005-06 season. 2. Broadcasters make limited use of multiplex capabilities (e.g., due to lack of cable carriage and/or production costs). 3. Consumers indifferent to free-to-air DTV for most of first decade of 21st century. 4. Lack of consumer interest dooms advertiser interest.
Spectrum Auctions **	1. Congress recognizes difficulty of shutting off analog in 2006 but makes it a policy priority to achieve turn off no later than Dec. 31, 2010; FCC instructed to facilitate. 2. Government continues pressure for auctions; channel 60-69 auctions occur in late 2001 or early 2002; broadcasters relocated prior to analog switch off in their DMA with incentives paid by auction winners. 3. Channel 52-59 auctioned in 2005 (three years late); relocation process similar to channel 60-69.	1. Channel 60-69 auctions occur in March 2002 but auction fervor subsides due to no realistic analog shut off plan; government frustrated at inability to generate auction revenues. 2. Channel 52-59 auctions deferred indefinitely 3. Broadcasters retain dual channels with digital channel largely underutilized; Government displeased. 4. Government imposes escalating spectrum use taxes	1. Channel 60-69 auctions deferred until 2003; bid revenue fails materially to meet CBO projections because bidders doubt spectrum will be vacated within business relevant timeframe. 2. Government suspends further spectrum auctions, resolves to offset auction revenue foregone with spectrum use taxes; such taxes based on hypothetical highest and best use (not actual use). 3. Industries that could use spectrum accommodated elsewhere and/or forced to adjust. 4. Situation persists until end of second decade (circa 2020).

* See next section for a discussion of must-carry issue.

** Spectrum auctions are relevant to DTV rollout because government agencies desirous of maximizing auction revenue have an incentive to take actions that support broadcasters vacating rapidly the auctioned or to-be-auctioned spectrum. Conversely, the government has an incentive to punish (i.e., tax) broadcasters if the perception is that the broadcast industry is delaying the auction process.

The CHAIRMAN. Thank you.
Dr. Hazlett.

STATEMENT OF THOMAS W. HAZLETT, PH.D., RESIDENT SCHOLAR, AMERICAN ENTERPRISE INSTITUTE FOR PUBLIC POLICY RESEARCH

Dr. HAZLETT. Thank you, Senator, and thanks for inviting me today.

To say that the transition to digital television is not going well is a bit like saying that Mikhail Gorbachev's perestroika is falling somewhat behind schedule. The disastrous failure of public policy is hidden only by lack of news coverage. This problem may be solved soon when the media comes to focus on this issue. The press, of course, loves a good train wreck.

Comparing the digital TV transition to perestroika is not gratuitous. The central planning mechanism at the heart of spectrum allocation in the United States through the Federal Communications Commission is a structure and restructuring process which looks at wireless telecommunications from the top down. This system is inefficient, unresponsive to consumer demand, and a huge barrier to entry for new technologies anxious to compete in the marketplace.

Recently at the Federal Communications Commission, a group of 37 economists expert in telecommunications policy, filed a comment urging liberalization in spectrum policy, and I would refer you to that document. It was signed by Nobel Laureate Ronald Coase, the immediate past chairman of the Council of Economic Advisors, Martin Bailey, and at least a half-a-dozen former Chief Economists of the Federal Communications Commission. It is available at the FCC or at the website of the AEI Brookings Joint Center for Regulatory Studies.

The history of DTV already reads like a Russian novel. It was born not in the laboratory, but on K Street, as an attempt by broadcast lobbyists to block land mobile services from getting access to UHF spectrum in the mid-1980s. High definition TV was the reason created for freezing any use of idle bandwidth, despite pressing demands for more wireless telephone competition.

Over a decade, technical standards were hammered out and complicated transition rules ordained. The result is technology adoption by committee. While a switchover date has been set in law, no one seriously believes that analog broadcasting will go dark in 2006. If they did, they would be buying digital TV sets. Of 100 million U.S. television households, only about 50,000 are equipped to receive digital off-air signals. What do consumers know that policymakers do not?

Well, this leads me to a brief discussion of today's policy choice, clamp down or loosen up. I think that discussion should start with this realization: Consumers correctly see high prices and major uncertainties. They don't see a killer app. They don't even see a modestly threatening app. The obvious solution, obvious to some, is to: (a) mandate digital compatibility for all newly sold TV sets in the United States; (b) mandate digital must-carry; and (c) move to a quick elimination of analog broadcasting. This approach concedes that only through brute policy guarantees will customers embrace digital TV.

Do not do it. As policy, this is ultra-high-risk. It lacks cross-checks from the marketplace and feedback from customers. It opposes costs on viewers, competitors and technology creators, who are eliminated from this analysis. In just one area, digital must-carry, significant costs may be imposed by soaking up valuable bandwidth on cable and satellite systems which distribute programming, only to distribute programming of little interest to customers.

By the way, satellite systems are even more impacted by the negative anti-competitive effects than are cable systems, and of course, it is the satellite television that is bringing competition to the multichannel video market.

In short, the brute policy approach puts us further down the path of industrial policy. It has a high probability of proving disastrous, forcing costs on the economy, while blocking more valuable wireless services. The superior solution lies in liberalization, quickly giving new competitors access to radio waves in the TV band. This can be achieved by giving broadcasters the freedom to offer extensive broadcast and non-broadcast service over both the new digital and old analog channels.

The FCC, however, should immediately allocate all unused TV bandwidth to new wireless licenses with broad flexibility. These would be called overlay rights. As only 13 TV stations broadcast in the typical U.S. market, even doubling such assignments with digital broadcasting leaves vast unused gaps in the 67 channels or 402 MHz allocated to TV. These overlay rights, these new rights would allow new users to access radio spectrum and should be assigned by competitive bidding. Winning bidders would then negotiate with current users, TV stations, to vacate their positions for a fee.

This would create additional bandwidth for new services such as 3G wireless and fixed wireless broadband access. It could also unleash vigorous competition to existing broadcasting, cable and satellite services. Thank you very much.

[The prepared statement of Dr. Hazlett follows:]

PREPARED STATEMENT OF THOMAS W. HAZLETT, PH.D., RESIDENT SCHOLAR, AMERICAN ENTERPRISE INSTITUTE FOR PUBLIC POLICY RESEARCH

Spectrum Allocation

1. To say that the transition to digital television is not going well is a bit like saying that Mikhail Gorbachev's *perestroika* is falling somewhat behind schedule. The disastrous failure of public policy is hidden only by lack of news coverage. This problem may be solved when the media come to focus on this issue in upcoming years. The press, of course, loves to cover a good train wreck.

2. Comparing the digital TV transition to *perestroika* is not gratuitous violence. The central planning at the heart of the spectrum allocation system leads the U.S. Government, through the Federal Communications Commission, to structure and re-structure wireless services from the top down. This system is inefficient, unresponsive to consumer demand, and a huge barrier to entry for new technologies anxious to compete in the marketplace. The consensus among policy economists is that the entire system is in need of substantial reforms allowing wireless bandwidth markets to emerge. In a February 2001 Comment filed with the FCC, 37 economists with expertise in telecommunications and public policy, including Nobel Laureate Ronald Coase, the immediate past chairman of the Council of Economic Advisors, Martin

Bailey, and six former FCC Chief Economists,[1] urged regulators to relax licensing rules such that existing operators can use spectrum flexibly and new competitors or technologies can challenge the status quo. This filing is available online: *http://www.aei.brookings.org/publications/related/fcc.pdf*

The Origins of Digital Television

3. Extending spectrum liberalization to the TV Band is easy at a theoretical level. Industrial policy is anti-competitive and ultimately anti-consumer. Competitive markets include far more nuanced information than FCC rulemakings, and are not biased by the political lobbying that pervades that process. When investors decide how to use radio spectrum they are careful to weigh the alternatives, searching for opportunities that may be unseen, undeveloped, or uncertain. They are calculating and relentless in discovering what is possible, what customers are willing to pay for, how much to invest in new technology, and how long to wait for new science.

4. At the specific level of implementation, these tradeoffs are crucial. Not only are digital TV sets, stations, and programming expensive to create, the use of bandwidth for digital TV crowds out potentially valuable services like cellular telephony, fixed wireless broadband, or 3G (mobile web services). Since the DTV transition has been mandated by FCC rulemakings, entrepreneurs have been prevented from attempting innovative ways to offer new services to the public.

5. The history of DTV already reads like a Russian novel. It was born not in the laboratory, but on K Street, an attempt by broadcasting lobbyists to block land mobile services from gaining access to UHF spectrum in the mid-1980s. High Definition TV was the reason created for freezing any use of idle bandwidth, despite pressing demands for more wireless telephone competition.

6. Over a decade, technical standards were hammered out and complicated transition rules ordained. The result is technology adoption by committee. While a switchover date has been set in law, no one seriously believes that analog broadcasting will go dark in 2006. If they did, they'd be buying digital TV sets. Yet, of 100 million U.S. TV households, only 50,000 are equipped to receive digital off-air signals.[2] What do consumers know that policymakers don't?

Clamp Down, or Loosen Up?

7. Consumers see high prices and major uncertainties about long-term adoption. They don't want to be stuck with expensive equipment that isn't needed and doesn't receive desirable programming. The seemingly obvious solution is to: (a) mandate digital compatibility for all newly sold TV sets (thereby getting economies of scale to kick in), (b) mandate digital must-carry, (c) eliminate analog broadcasts in 2006. This approach concedes that only through brute policy guarantees will customers embrace digital TV.

8. Don't do it. As policy, this is the ultra-high-risk approach. It assumes that the digital television transition, as mapped out, is the one and true path to consumer satisfaction. And it does so without cross-checks from the marketplace, feedback from customers. Costs to viewers, competitors, and technology creators are eliminated from the analysis. In just one area—digital must-carry—these costs may be terribly high, soaking up valuable bandwidth on cable and satellite systems to distribute programming of little interest to customers.[3] In short, this approach puts us further down the path of industrial policy. It has a high probability of proving disastrous, forcing costs on the economy while blocking more valuable services.

9. The superior solution lies in liberalization, quickly giving new competitors access to radio waves in the TV Band. This can be achieved by giving broadcasters the freedom to offer extensive broadcast and non-broadcast service over both their new (digital) and old (analog) channels. The FCC should immediately allocate all unused TV band airspace to new wireless licenses with broad flexibility. As only 13 analog stations broadcast in the typical U.S. market, even doubling such assignments with digital broadcasting leaves great unused gaps in the 67 channels (402

---

[1] The list of signatories includes: Martin Neil Baily, Jonathan Baker, Timothy Bresnahan, Ronald Coase, Peter Cramton, Robert W. Crandall, Richard Gilbert, Shane Greenstein, Robert W. Hahn, Robert Hall, Barry Harris, Robert Harris, Jerry A. Hausman, Thomas W. Hazlett, Andrew Joskow, Alfred E. Kahn, Michael Katz, Robert E. Litan, Paul Milgrom, Roger G. Noll, Janusz Ordover, Bruce Owen, Michael Riordan, William Rogerson, Gregory Rosston, Daniel L. Rubinfeld, David Salant, Richard L. Schmalensee, Marius Schwartz, Howard Shelanski, J. Gregory Sidak, Pablo Spiller, David Teece, Michael Topper, Hal Varian, Leonard Waverman and Lawrence J. White.

[2] Christopher Stern, *Mixed Signals, Broadcasters' Promise of a Digital TV Age has Not Been Met, And Now Congress Is Having Second Thoughts About Its Role*, Washington Post (Dec. 17, 2000), H1.

[3] See Thomas W. Hazlett, *Digitizing Must-Carry Under Turner* v. *FCC* (1997), *http://www.aei.org/ra/rahazl1.pdf.*).

MHz) allocated to the TV band. These overlay rights would allow new users to access radio spectrum, and should be assigned by competitive bidding. Winning bidders would then negotiate with current users (TV stations) to vacate their positions for a fee. This will create additional bandwidth for new services, such as 3G wireless. It could also unleash vigorous competition to existing broadcasting, cable and satellite services.[4]

The CHAIRMAN. Thank you, Dr. Hazlett.

Dr. Cooper, in the past, your organization supported the broadcasters efforts to have the government guarantee carriage of all broadcast stations on cable systems. Now you support a full free market auction of spectrum that the broadcasters claim they should have for free.

Why have you moved to this free market approach?

Dr. COOPER. Well, our view of the spectrum has to do with the alternative uses that are available here, and what we have learned in the past half decade, particularly with the statistics I gave you, is that there is an immense potential for the use of that spectrum that has a great deal of value to the public.

We never believed that spectrum should be given away for free. We always were supposed to get compensated for it in the past through public interest obligations. Going forward, we think the best way to extract the public's value for the public's resource is to mine it in terms of its alternative uses, and make sure those funds remain, flow back to the public.

And in a certain sense, I would disagree with the suggestions that were made by the last two speakers who want to allow the broadcasters to sublet something that they never rented. They have never paid for that stuff, and so creating that secondary market, we want those dollars for that sublet to end up back in the public's pocket, and we want it to be used again in the public interest ways we have identified.

But clearly, you have to recognize the value of this real estate which is owned by the public, and that is the fundamental driving force in our change.

The CHAIRMAN. Thank you.

Mr. Gattuso, one of the penalties of coming before this Committee is that you have been here before. In March 1996, you testified before this Committee on spectrum policy. During the hearing you stated, "spectrum should be treated more like other resources in society, giving its users the ability and incentive to put it to its best use."

Do you believe the spectrum is being put to its best use, and what are the consequences as a result of this spectrum being given away to broadcasters for free?

Mr. GATTUSO. Well, first, I think we are dealing both in this field as in most other fields, especially in any field of technology and any field of great uncertainty. Government policymakers as individuals do not know what the best use is, and that is why we have to set up processes, market-based processes, to determine that.

My own personal view is that the spectrum is not currently being used for its best use. I see, as evidenced by wireless auctions in other areas, a huge amount of value in other uses, and I am very

[4] For further elaboration, see my "Essay on Airwave Allocation Policy," forthcoming in the Harvard Journal of Law & Technology: *http://www.aei.brookings.org/publications/working/working-01-02.pdf.*

very worried that we are holding back 12 MHz of spectrum, or at least 6 of the 12, at the expense of these much more valuable uses.

The CHAIRMAN. Dr. Kraemer, your testimony indicates you would take more of a regulatory approach to solving the transition to DTV. You mentioned that the FCC should require that all new television sets that are sold should include the capability to receive digital signals. Currently, digital tuners cost $500; it is estimated they will still be as high as $300 by 2003.

By requiring digital tuners that will double or triple the price, do you really believe that that is in the best interest of consumers to pass on such a mandate?

Dr. KRAEMER. Senator, I do not think it will double or triple the price. The reality is what you are looking at is the chip set that essentially does the conversion, and effectively you are talking about chips. In the end, chips cost less than a dollar. If the chip manufacturers know that everybody must do it as of 1/1/04, they will essentially create a chip that in a single chip takes both NTSC, which is over-the-air analog, and ATSC, over-the-air digital, and put it into a single chip. And for that matter, they may even put into that chip the cable standard, which is QAM.

So that, in the end, a great deal of what we are talking about here will be taking place at the chip level in the set, and there is a large room for technology to be effective here.

The CHAIRMAN. Dr. Hazlett, in your testimony you argue that broadcasters should be given the freedom to offer extensive broadcast and non-broadcast service over both their digital and analog channels. But I gather you do not believe that the U.S. taxpayer should continue to pick up the tab for this flexibility.

To what extent should the broadcasters continue to be able to use the spectrum for free if they are going to then turn around and use it as a profit-making mechanism?

Dr. HAZLETT. The question of whether or not to auction these licenses was a very lively question, as I know that the Senator recalls in some detail, and I was also here in March 1996—call it March madness—and was testifying in favor of auctions at that time, and I am glad that more people now are coming to the auction view.

But the problem is, today you have a consumer welfare train wreck on your hands. The licenses have gone out. Broadcasters are starting to invest in these new technologies, and some consumers are actually starting to invest by buying these expensive, in fact, very expensive TV sets.

The concern should be how to get new services, competitive services, to consumers in the marketplace. If you are going to worry about mistakes that were made in the past, you will be here having a hearing in 5 years, 10 years, 15 years, talking about how the digital transition for television is going. The thing to do now to get at the broadcasters and to institute some equity is to introduce to the broadcasters. That is the way to get at the value of the licenses, and in fact, produce additional economic activity, competition, lower prices, and in fact, tax revenues for the U.S. Treasury. That is the way to get equity.

The CHAIRMAN. But they use either analog or additional digital spectrum for other uses; OK?

Dr. HAZLETT. Right.

The CHAIRMAN. Then they are competing with people like wireless, who have to pay for their spectrum.

Dr. HAZLETT. Right.

The CHAIRMAN. In the case of the last spectrum, $17—what, auctioned $17 billion. How do you compete? How do people paying for their spectrum compete with people who have it for free?

Dr. HAZLETT. Well, I was actually one of those who said back in the 1980s—and I can give you the citations—that cellular telephone licenses should have been auctioned. They were not auctioned, but those cellular telephone licensees compete head-to-head with PCS licensees. One is auctioned, one has not been. So you have got these inequities. There are inequities everywhere.

Now if you are going to spend time and political capital figuring out the inequities, not only are you not going to solve the inequities, I can guarantee you that——

The CHAIRMAN. I am not trying to solve the inequities, I am trying to at least give some kind of competition capability. If you get a baseball team for free, and I pay $700 million for it, it is very hard for us to compete for the players. I mean, I am not an economist, but look, I am not trying to right the inequities, but if you have an inequity that gives one of the competitors a dramatic advantage, then you have to do something to level the playing field.

Dr. HAZLETT. Well, the dramatic advantage is already sunk, and in fact, the taxpayers——

The CHAIRMAN. Sure it is.

Dr. HAZLETT. The shareholders that got the advantage of that auctioned it off. They have gone, they have sold. You cannot even get the people who got the advantage of those free licenses that were awarded in 1997. But right now, you hurt consumers by delaying for years of even shorter periods, by delaying new services that could compete with the broadcasters in the TV band.

So the real solution to the problem that has been created because of policies that were much to aligned toward industrial policy——

The CHAIRMAN. Do you believe the analog spectrum should be given back?

Dr. HAZLETT. If you want to put a trip-wire on that, I do not object to that. But the fact is, if there are millions of people that do get analog services, that looks like a very popular service. And if you are going to put all your chips on this transition that analog is going to go back by 2006, obviously there is a train wreck right there, customers are not going to get TV. That is why nobody seriously thinks that there is going to be a 2006 switchover.

So you have to respect what consumers have invested in and not kill—I mean, the objective of this is to help customers. And you cannot lose sight of that because you are worried that the broadcasters have some advantages, and they have great advantages. I have been quite concerned about this in the political process.

But the fact is, they have already won that war. If you continue to fight that war, you are going to lose this one. This one is for consumers right now that want digital services like 3G and fixed wireless.

The CHAIRMAN. I am sure that our dialog has stimulated some response from our other members, so I will just go right down the list. Dr. Cooper, and then Dr. Kraemer and Mr. Gattuso.

Dr. COOPER. Well, one of the key things from our point of view is no more concessions. Dr. Hazlett is talking about a sunk cost and he is trying to find a workaround around that. The first answer is not to make more concessions or not to reach back into the consumer's pocket.

Now if you can deliver that all-purpose tuner for a buck, we will not complain about it, but no one believes you can. The dollar chip that is going to receive all signals comes after an awful lot of front-end fixed costs that the manufacturers are going to try to recover.

It may well be if public policy needs to do that, the first thing you should do is make sure the public does not pay, so that the broadcasters maybe should pony up the development cost for that tuner, so that it does not end up in increasing the cost of my TV set. One possibility.

You can tax one group to make sure that you accomplish your industrial policy. Frankly, we would rather go the opposite way. Subject this resource to a market test as soon as possible, reallocate those licenses according to their highest value, and make no more concessions to folks who have been given the most important asset, investment asset, in getting us to the digital age.

The CHAIRMAN. We will go in order. Mr. Gattuso and then Dr. Kraemer.

Mr. GATTUSO. I think the most important thing is, as Tom Hazlett said, is to help consumers, and to make this spectrum available for its most valuable uses. I would love to make the broadcasters pay for what they got for free. That potentially can be done, as mentioned in my testimony, by setting a firm cutoff date. I do not know how politically possible that is. But that is something that economically would be fine.

If that is not possible, it is much better to take the steps that are necessary to insure that this spectrum is moved to its most valuable uses so consumers are helped. Even if broadcasters do not end up paying in the end, consumers should be the first priority.

The CHAIRMAN. Dr. Kraemer.

Dr. KRAEMER. If you want to accomplish a quick transition so that broadcasters start out with one channel, now have two, and go back to one, well then, you really have to focus on what your leverage points are around making that happen.

A second point I would make is that we have evolved a devil theory of the broadcasters, which simplifies the debate, but which is misleading. The reality is there is a very complex value chain here. Multiple manufacturers, none of whom are U.S.-based, manufacture sets. You have networks that do the programming, and they buy it from a whole series of studios, many of which are not integrated and have nothing to do with the networks. And then you have broadcasters, some of whom have three stations, some of whom have 50. A very diverse industry. And when you just say the broadcasters, you make it very difficult to pin it down.

The third thing is with these spectrum auctions, you may need to look at the economics. The more spectrum you make available, the less valuable it becomes; therefore, the auction prices go down.

Another thing is, you cannot chop up the spectrum. You do not do channel 7 in Philadelphia and channel 13 in Phoenix. Wireless operators want a single frequency nationwide; that is why they bid the money.

So you really need to get out a fact set around some of these issues.

The CHAIRMAN. Thank you, Dr. Kraemer. I would mention that as more spectrum has become available, more use of the spectrum has occurred as well, which has actually driven up the value of the spectrum in a rather dramatic fashion. The next wave spectrum which went for $4.5 million originally, a billion originally, recently was auctioned off for $17-some billion. Quite a remarkable increase in value, but I think your point is well made.

And I thank the witnesses. You have been very helpful.

Senator Burns.

Senator BURNS. Thank you, Mr. Chairman. I only have a couple of questions and I am just going to let you all react to it.

When we were wrestling with this problem back in 1996, and in fact, whenever we decided we would sell spectrum at auction, should we have sold the digital spectrum at auction and let the broadcasters invest if they wanted to, and still allow them to retain the ownership in the analog spectrum? In other words, should we have sold it to the highest bidder and then let the broadcasters, if you want to stay in the covered wagon without going on the rails, why, we would stay in the analog business. Should we have done that? Your reaction.

Dr. Cooper, now I want to tell you what you sound like. I know you do not want meanness, but I want to tell you what you sounded like a minute ago. You want the consumers to pay nothing for the service, yet you want them to collect the money. In other words, you want their cake and eat it too, and I think the consumer has a responsibility in this also, and the responsibility is that I think you cannot ask the broadcasters to pull up both ends of the wagon. That is what you sound like.

I just want to answer that question though. What would happen if we had just sold that spectrum and said, it is going to be designed for digital or high definition television, and anybody can buy it. Whether you are in the broadcasting business or not, here is your chance to get into the broadcasting business on the cutting edge, and not even bother about the requirement of returning the analog spectrum.

Dr. COOPER. In point of fact, the consumer buys the set and the consumer watches the advertising, so they pay at least with their opportunity, cost and time, and that drives the TV industry, so they do not get it for free.

If you had sold the spectrum, which would have monetized the public value there, and remember, this is public money in our view. If you had sold that spectrum, you would have a lot more programming, because——

Senator BURNS. Let me get it. It is not public money until I spend it.

Dr. COOPER. Well, it is not public money until it comes in the Treasury and we want to put it in a special place and use it for specific purposes, which was have outlined. But it is clearly a public resource owned by the public which has been rented at no price, given away. If you had sold it and they had a real capital cost on their books that they had paid for that asset, they would be getting a lot more value out of it. It is only because they had a free good that they have not had to exploit it.

And so, they have got the best bottom land in the county and they do not have to plant any crops, because it does not cost them anything. And they are still trying to figure out what they want to do with it. And so from our point of view, if they had paid for it, we would be getting a lot more value out of it, we would be buying the digital TV sets because there would be programming out there.

And so we do not get that—we do not pay nothing. Consumers pay for the TV and they watch the advertising.

Senator BURNS. Well then, should that have—and not the requirement of turning back the analog spectrum, that end of it. We are just saying we should have put it on the open market and sold it, without any attachment to the analog spectrum.

Dr. COOPER. You should have rented it, and we like to make the point that we own it in perpetuity, you rent it for a period of time, you do not own it forever, and when the lease runs out we get to reevaluate it. We do not want to lose control over it permanently, so as the value rises we can—just like a piece of property.

Senator BURNS. Do you want to comment on that, Dr. Kraemer?

Dr. KRAEMER. Let me just respond to your original question. Senator, the issue would be, after you auction this, do you want free-to-air television. In other words, if you put this up for auction and the bidders paid for it, they are not going to continue free-to-air television. They are going to have some type of subscription wireless service, and that is what you would have gotten.

Now if that is what you want, that is fine. The public treasury would have gotten money, and if people wanted to see the digital programming, they would have paid for it. What you would have had was wireless cable. The issue is, as I understood the debate, you wanted to maintain free-to-air television.

Senator BURNS. Let us say that we just put the requirement that you have to have free over-the-air television. That is the only requirement we made. You had to be a broadcasting free over-the-air broadcaster to buy it. That is what you are going to use the spectrum.

Dr. KRAEMER. Well, you probably would have reduced the value. The other thing you might have done is kill analog, because the buyer would have been General Electric or somebody like that, who basically would have been able to, at least in theory, cut loose from the analog side of it and the local broadcasting, to create an integrated national digital network.

Senator BURNS. OK.

Yes, sir, Dr. Hazlett.

Dr. HAZLETT. You have actually asked a very interesting and a very difficult question, and it sort of gets to the reason of why you see these estimates that these digital TV licenses are worth between $12 and $70 billion. That is a substantial range. In fact, the 95 percent confidence interval is probably wider than that. The FCC does not really know what these licenses would have garnered, particularly under the conditions that you suggest, where analog stays where it is, and there is just a new license called digital TV, how much will you pay.

Now what the evidence is now, if this was the anticipation, that there is very little consumer demand evident in the marketplace, given all the transitional difficulties, that would have been reflected in very low bids for the licenses.

But let me further suggest something, that there is a little bit of a semantic problem with all these discussions about spectrum auctions and giving away spectrum to the broadcasters. At the heart of the problem is that broadcasters were not given spectrum. The broadcasters were given TV licenses and they were forced to deliver a product with the license. That is what they can do with the spectrum. The FCC is the one that has the spectrum. They allocated it to TV, specifically digital TV, and they set the rules.

Now if the spectrum is worth a lot more providing fixed wireless broadband or 3G wireless mobile service, that cannot be done on a TV license, barring some future policy change at the FCC. So the real problem is that these licenses lock in regulated uses.

So now you have a situation where you did have this quid pro quo, provide digital TV and we will give you this free license, where we have locked into that technology by virtue of that transaction, and other services that are much more potentially valuable use of the spectrum cannot get access to those radio waves. That is our basic policy conundrum and that is why there has to be substantial liberalization of this entire industrial policy approach to really deliver value to customers.

Senator BURNS. Mr. Gattuso.

Mr. GATTUSO. You need to create an opportunity cost. I think that as I testified 5 years ago, an auction would have been the best way to go and that would have helped insure the people who were getting the licenses have good plans for it, that they are willing to put their money down on that, behind the fact that they could use this resource.

But that would not have been enough. As Tom just mentioned, if you do not have that opportunity cost, if you do not have the possibility of using the frequencies for an alternative use, you are not insuring that it will be used for the highest and best use. That is why I supported plans for negotiations and allowing other wireless providers to use the spectrum. Without that, you still are locking in a potentially lower value use. I say potentially, we do not know for sure now, but you do need to find out.

The CHAIRMAN. Senator Fitzgerald.

Senator FITZGERALD. Well, it looks like we cannot go back and change what we did in 1997, and obviously we gave away some very valuable spectrum to the digital broadcasters. My concern now is that we salvage this situation and that's why I want to focus on whether we try to salvage this situation by mandating, as Dr. Kraemer suggested, that new TV sets contain the chip so that they could receive the digital broadcast, and then require the broadcasters to give back their analog spectrum after a reasonable period of time, and then we could reallocate the analog spectrum to its highest and best use, hopefully through an auction.

Now, Dr. Hazlett, you were opposed to that because that was further going down the road to the industrial policy. Does it not seem,

Dr. Hazlett, in order so that this does not become a bigger boondoggle than we already know it is, that we actually have to go the heavy-handed step of mandating that new television sets carry the chip, or else these broadcasters will be squatting on both the digital and analog spectrum for as long as we can foresee, probably until 2020, as Dr. Kraemer suggested.

Dr. HAZLETT. No. The fact is that these kinds of—this is exactly why industrial policy ends in these Byzantine regulatory structures where you are fixing—you know, years later you are fixing problems. You never know how this whole thing started, and in fact, right now, we do not remember this whole thing started because there was a dispute at the FCC in 1985 about whether or not to give more UHF TV spectrum to cellular because it was not being used by TV, and that led to high definition. High definition is long gone, now we are talking about digital TV, now we are talking about mandating must-carry.

The fact is that all these requirements are going to be very expensive. Now if it really does cost a dollar for the chips, the chips will be provided by the market. The problem is, it is not going to cost—according to the set manufacturers that are holding off on this thing—it is not going to be a trivial cost. If it is $100, that is $10 billion to equip 100 million TV sets. That is real money, and that is only one TV set per television household.

If you start mandating these things like must-carry, you drive off CASPIAN and——

Senator FITZGERALD. OK. But if we don't do that aren't we going to see that broadcasters continue to retain their analog spectrum and have the digital TV, and we are wasting a lot of spectrum?

Dr. HAZLETT. Well, you are wasting spectrum, not because the broadcaster is holding it, you are wasting it because of the rules that lock in inefficient use. That is why the suggestion that I made was to liberalize that, allow the broadcasters to compete in these other markets, and to allow others, new entry to compete in these other markets. That can be done through these overlay rights that are auctioned off, compete head-to-head with the broadcasters in addition to all the competitors that are out there in some of these spaces today. And new competitors through additional spectrum liberalization should be invited in a general policy which is very favorable to consumers in efficiency.

But you cannot undo those old mistakes. You know, I was here to argue the other way. But the fact is that you have to deal now with what is in the consumers' interest in going forward. If you go down the road to intensify the industrial policy, you are going to make a huge gamble with consumers' dollars.

Senator FITZGERALD. Mr. Gattuso, I know you have had your——

Mr. GATTUSO. I think the basic premise we need to follow here is that two wrongs do not make a right. We did not auction the digital spectrum, we did not put in proper service rules. Further regulation will only make the problem worse. I would not want to be in the situation where responding to these constituent calls when the cost of televisions does do go up. I know people said that will not happen, it could happen. I would not want to be taking the calls explaining why because we gave broadcasters $X billion worth of free spectrum now they have to pay more for their televisions.

Senator FITZGERALD. How do we get the analog spectrum back, though?

Mr. GATTUSO. Everyone seems to assume today that there is a chicken and egg problem that is unsolvable, and that chicken and egg problems cannot be overcome. Chickens and eggs exist; I have seen them. If it was not for the market's ability to get over these chicken and egg problems, you would not have CD players, you would not have DVD today, you would not have VCRs, you would not have virtually anything.

In the marketplace, when there is a good product with consumer demand, there are ways to get these problems. There has been a lot of good academic work on this. Stan Liebowitz at the University of Texas, who has done a lot of work on network effects, finds out that good products get over this problem, even though it does at first look unsolvable.

Senator FITZGERALD. Dr. Cooper.

Dr. COOPER. Mr. Fitzgerald, if you want to go down the route of—basically you are looking for a penetration price on the tuner. You want to get the tuners out there in the world. And I have suggested that you should not ask the public to pay for that. The question is, who benefits from this network effect? We have just heard about the chicken and egg problem and network effect.

The answer is, the broadcasters are the primary beneficiaries. Set manufacturers cannot possibly benefit from this because they simply sell a piece of hardware in the middle of the network, right? The guys who benefit from it are the ones who get the eyeballs. And they told you, we may have 6 times as many eyeballs in the sense that we have many more channels in that one space.

So they are the folks who should, in fact, be willing to engage in what is called penetration pricing. They ought to be willing to price below cost on this network element, because when the network grows, they get the benefit, in addition to which they have already received the benefit of having the asset for free.

So if you are contemplating forcing tuners on the public, the answer is the public should be held harmless, the broadcasters should be the one to subsidize the front end.

Senator FITZGERALD. Make them buy the tuners?

Dr. COOPER. Make them pay for the tuners. And of course, they will have a real interest in getting those tuners out there in the cheapest manner possible, and they are the beneficiaries of network effects.

Senator FITZGERALD. I agree with that, but just realistically, I do not think that will ever happen. Even though I might vote for something like that, that obviously is not going to happen from what I can tell right now.

Dr. COOPER. Well, we will pat you on the back for trying.

Senator FITZGERALD. All right. Well, thank you all.

The CHAIRMAN. Thank you very much. It has been a very good hearing. Is that about it?

Senator BURNS. You can go as long as you want, but I am going to dinner.

The CHAIRMAN. Yes, there you go. Well, thank you all very much.

Thank you, gentlemen, for coming today, and your testimony, and I know there will be other members of this Committee who will probably require information. If you get questions, please respond to the senders and to the Committee.

Thank you for coming. The record remains open.

[The hearing adjourned at 11:55 a.m.]

APPENDIX

PREPARED STATEMENT OF HON. ERNEST F. HOLLINGS, U.S. SENATOR FROM SOUTH CAROLINA

Today's hearing examines issues related to the transition of the broadcast industry from analog to digital service. As early as 1986, broadcasters had begun advocating that they needed to transition to "high definition" television. They worked to have the FCC begin the necessary regulatory process for the transition and to obtain legislation from Congress with respect to the transition. It has taken a number of years to develop high definition television, and at times, the progress has been difficult.

Today's hearing will certainly provide us with a better understanding of where we are in the process and the remaining issues that need to be resolved in order for the transition to move forward. Indeed, a thorny issue which must be resolved is that of "must-carry." During the transition broadcasters expect to be transmitting both analog and digital signals and expect both signals to be carried by cable operators. Broadcasters also have argued that cable operators must carry all of their free digital programming. However, the cable industry opposes the broadcasters' position on these issues.

In order to move forward, the FCC must conclude its review of these issues quickly and in a manner that ensures a successful conversion to digital television. This means ensuring that consumers can receive their broadcast signals during the transition and cable networks are not unfairly displaced in the process.

Another issue that also needs to be resolved in order for this transition to move forward is how to protect digital content from being illegally misappropriated while also protecting the rights of consumers to use and record programming. I suspect that making digital programming available to consumers will stir demand for programming and equipment and ultimately, speed the transition. Therefore, I encourage the parties involved to resolve this matter.

The additional issues that also must be addressed include the buildout and upgrading of broadcast stations and the availability of digital television equipment so that consumers can see high definition programming.

In the end, in order to obtain the rewards of digital television, everyone must continue to work together to resolve the difficult issues that still exist. Ultimately, the transition to digital television will be a success if consumers have greater choices at affordable prices.

I welcome the witnesses and look forward to hearing their testimony.

PREPARED STATEMENT OF RICHARD M. LEWIS, SENIOR VICE PRESIDENT, RESEARCH AND TECHNOLOGY, ZENITH ELECTRONICS CORP.

As a long-time participant in the digital television (DTV) transition, Zenith Electronics Corporation appreciates the opportunity to submit this statement for the record discussing where we are today, the challenges we face, and the steps that are necessary to complete the transition to DTV.

We are pleased to report that DTV sales are growing and customer satisfaction levels are high with these products. In fact, according to the Consumer Electronics Association (CEA), approximately 687,000 DTV displays and receivers were sold in 2000, accounting for $1.4 billion in consumer spending. These numbers represent a sevenfold increase over the previous year. Looking forward, CEA estimates that unit sales of DTV products will grow 80 percent in 2001, with consumer investment climbing to $2.1 billion.

Not only do these numbers exceed CEA's initial projections, but they also compare favorably with previous blockbuster consumer electronics product introductions. Annual unit sales growth and dollar sales for DTV during its first 4 years on the market is projected to surpass those of computers, VCRs, CD players, and color TVs. This consumer interest is due to the wide variety of DTV products currently on the market. Over two dozen manufacturers have introduced more than 200 different DTV products, which are being sold at more than a thousand retail locations around the country. Availability increases every day as prices come down, more models are introduced and new retailers begin stocking DTV.

Best of all, consumer interest and satisfaction with DTV continues to rise. When consumers see the extraordinary sound and video experience offered by DTV, they want it—and today's analog television never looks the same again. Consumers are buying DTV even in those markets where broadcast programming is limited or unavailable. Americans are finding that digital and high-definition displays enhance the analog TV experience, and provide the best display for DVD and other pre-recorded content.

While these facts show that DTV momentum is growing, we at Zenith do not suggest that the DTV transition has advanced as quickly as needed or as far as possible. One issue that has previously impeded the DTV transition has been the debate over the DTV transmission standard. Fortunately, that matter now is resolved.

As one of the original developers of DTV technology in general and inventor of the vestigial sideband (VSB) transmission system in particular, we at Zenith are understandably pleased by recent reaffirmations of the ATSC standard. In January of this year, the Federal Communications Commission (FCC) reiterated its long-standing support for VSB and stated there is absolutely no reason to revisit the DTV standards issue. Also in January, the boards of directors of the nation's leading broadcast trade groups, the National Association of Broadcasters (NAB) and the Association for Maximum Service Television (MSTV), voted overwhelmingly to stay the course on VSB modulation and reject a European alternative. Repeated testing by the FCC's own Office of Engineering and Technology shows that the current 8-VSB transmission standard should be retained. All the evidence confirms that the 8-VSB standard is the correct standard for use in the United States, and with this debate resolved, manufacturers, broadcasters and consumers have the certainty they need to invest in further DTV enhancements.

As proof of our industry's focus on meeting market needs and willingness to cooperate, we continue to explore possible enhancements in the ATSC DTV standard to address broadcasters' changing needs. Receiver manufacturers and chipmaking labs are moving forward aggressively with improved designs for standard applications as well as proposing extensions to provide additional capabilities and flexibility. Because the ATSC standard was designed to offer plenty of "headroom," we are confident that a number of VSB enhancements will be adopted in the near term. Zenith has two such enhancements under development: E-VSB (Enhanced VSB), which would break the 19.4 megabit-per-second bitstream into two parts, one for regular HDTV and another for more robust applications, such as datacasting, and M-VSB, which would provide a mobile solution if broadcasters decide they need such applications.

We urge you to consider the remaining roadblocks to widespread DTV acceptance. In our view, there are four such impediments: (1) the lack of compelling digital content; (2) affordability of consumer equipment; (3) cable carriage and interoperability issues, and (4) the digital copyright situation. Not surprisingly, these issues cut across multiple industries—broadcast, consumer electronics, cable and programming—and therefore pose some thorny challenges for both the private sector and U.S. policymakers.

Broadcaster Activities

U.S. broadcasters have made impressive investments in DTV transmission equipment. With more than 180 stations currently broadcasting a DTV signal, the industry is far outpacing the DTV transition timetable established by the FCC. Most broadcasters are meeting, if not exceeding, their obligations to begin DTV service. While some stations have encountered tower siting and construction problems, the majority of major network affiliates in the 30 largest media markets are broadcasting in digital. Special credit goes to the growing number of stations in smaller markets—such as Quincy, Illinois (market number 161) and Salisbury, Maryland (number 162)—that have begun DTV broadcasting well in advance of the government-mandated schedule.

At the risk of being labeled optimists, we at Zenith continue to believe that the 2006 deadline for effecting the digital conversion remains theoretically achievable, assuming that the key industries come together to reach agreement on the issues identified above. The real barrier to this timetable is that the transition has not yet captured the hearts and minds of American consumers. For all of us to succeed, consumers need a reason and the means to adopt these new technologies. Without compelling content (whether HDTV, datacasting or some new application), DTV will not flourish. Without equipment that the average consumer can afford, DTV will become a footnote in the digital age. The lack of access to DTV signals, whether over the air or through cable, renders all other issues irrelevant.

Importance of Digital Programming

Broadcasters can do their part in the all-important area of digital programming, a critical element in the overall DTV equation. Despite the leadership provided by CBS (which accounts for the lion's share of HDTV programming), the major commercial networks have yet to feed, let alone originate, their fair share of digital content. In addition to its commitment to prime-time programs in HD, CBS has offered an unequaled amount of HDTV sports programming, including the AFC playoffs, the Super Bowl, the Masters and the NCAA Final Four.

Without the efforts of CBS, PBS and a small number of independents like WRAL-TV in Raleigh, North Carolina, that have produced and televised a number of exceptional programs (and pushed the envelope on data broadcasting), the early DTV purchaser would have virtually nothing to watch in true HDTV. In fact, absent far greater amounts of compelling digital content, consumers will have little incentive to make the investment in DTV equipment, especially at today's prices.

Besides HDTV, innovative applications of multiple standard-definition television (SDTV) and datacasting may also prove compelling and help drive the DTV market. While we support broadcasters' efforts to provide supplementary and ancillary services such as datacasting, these efforts must not come at the expense of their primary obligation—to provide consumers the opportunity to experience high-quality HDTV programming.

Affordable Consumer Products

Given the meager amount of digital programming, it is remarkable that initial sales of consumer DTV equipment have posted such respectable numbers. Predictions by CEA that sales of DTV equipment this year will exceed one million units compares favorably with the sales curve of color TV, for example, which needed a full decade to reach sales of one million units annually. While some may point to the small number of tuners sold to date as an indication of DTV's failure, the reality is that the high number of sales of HDTV displays proves consumers want digital television. Today's consumers are very sophisticated and will not pay extra for items requiring content that is not available. Increased HDTV broadcast programming or other digital content will give them the reason to spend the extra money for a tuner or integrated set.

Over the last 2 years, receiver and display prices have been reduced by nearly half. This decline in the prices of DTV sets is in line with the 44 percent decline seen for DVD players, and much more rapid than the initial price declines of products like CD players, VCRs and large screen analog TVs. In addition, a wide variety of set-top boxes in the $600 range have been introduced, including boxes that incorporate reception for satellite and over-the-air DTV signals as well as analog signals. With the certainty provided by broadcasters' reaffirmation of the 8-VSB standard, CEA expects to see a strong upsurge in sales of set-top receivers this year.

Zenith is doing its part to offer consumers a wide array of quality DTV products at affordable prices. At the Consumer Electronics Show in January of this year, Zenith expanded its DTV line to include not only new widescreen (16:9) integrated rear-projection HDTVs and 16:9 HDTV monitors (plasma, direct-view, LCD and projection), but also the industry's first fully integrated digital TV set priced below $1,000. Consumer electronics is an intensely competitive business, and history suggests that it won't be long before demand for DTVs explodes, critical mass and production efficiencies are achieved, prices fall to even more affordable levels, and the product begins to earn mass-market acceptance. Once consumers experience the crisp images and theater-quality sound of DTV, they'll never go back to analog, particularly as DTV products become more affordable and available.

Cable and Content Producer Cooperation Needed

With some 70 percent of all U.S. TV households receiving their local, over-the-air stations via cable, the cable industry also needs to be on board if the DTV transition is to succeed. While some headway has been made on the issue of compatibility between cable equipment and consumer electronics products, we are concerned by the cable industry's slow pace in devising standards that will allow DTVs to connect to cable systems.

The cable issue proving to be even more difficult is the digital must-carry controversy. Ignoring the pleas of broadcasters that cable companies should be required to carry each station's analog *and* digital signals during the transition, last month the FCC ruled preliminarily that cable operators must carry only one or the other signal. Logically, if a broadcaster is only upconverting analog content, it is hard to understand why a cable company should be required to carry two versions of the same content. On the other hand, if the broadcaster is providing a different program stream, HDTV or SDTV with additional data content, the consumer is provided with benefit beyond analog television. In this case it is hard to see why the cable company should not provide the full, undiluted benefits of broadcast-quality DTV or true digital HDTV. Failure to resolve this must-carry issue presents a huge potential barrier to the DTV transition.

Digital copyright concerns must also be addressed and resolved if DTV is to succeed in the marketplace. Deeply troubled by the Internet music phenomenon and fearing the "Napsterization" of movies, the Motion Picture Association of America is expected to advocate severe limits on digital video copying. While no one condones the crime of video piracy, the Supreme Court held in 1984 that consumers have a right to make copies of TV programs, including movies, provided they are for "personal, non-commercial use." A reasonable balance must be struck between the legitimate concerns of content owners, on the one hand, and the well-established principle of "fair use" on the other. The digital age complicates, but should not fundamentally alter, the traditional fair use rights of consumers. As with other delivery media, any attempt on the part of copyright holders to deny consumers an over-the-air movie or other program, or degrade its transmission quality, obviates one of the primary reasons for buying widescreen HDTV: to enjoy movies the way they were intended to be seen.

Conclusion

No one said that the DTV transition would be easy. But there is momentum. We are pleased that DTV product sales thus far are well ahead of our industry's projections and that the number of DTV stations is growing as well. Now, manufacturers, broadcasters, cable operators and content producers must work together to formulate lasting solutions that will allow this fledgling medium to succeed. With the transmission standards debate behind us, we must resolve these few remaining issues and make DTV happen for the benefit of American consumers, who in the end will judge the true success of the DTV transition.

PREPARED STATEMENT OF RALPH M. OAKLEY, VICE PRESIDENT AND BROADCAST GROUP HEAD, QUINCY NEWSPAPERS, INC.

Quincy Newspapers, Inc. (QNI) is a privately held, family owned media company headquartered in Quincy, Illinois. We have provided service to the public through five generations. QNI operates six network-affiliated television stations in medium- and small-sized markets in the Midwest. As is the case with many other television broadcasters, QNI's ability to continue to serve its audience effectively in the future will be impacted significantly by the transition to digital television.[1] We submit this statement in the hearing today to make our views known on this extremely important matter.

It is important to State at the outset that QNI is committed to digital television and is working hard on the transition. Of QNI's stations, two have already completed construction of digital facilities in advance of their required May 2002 deadline and are presently on-air digitally. In this statement we report on our progress and inform the Committee of the practical challenges we face in our communities in making this transition.

A. Background on QNI

We first provide background on QNI. The company has long held media interests, as the roots of the company date back to 1835 with our newspaper, The Quincy Herald-Whig, being the lineal descendant of the Illinois Bounty Register. The ownership of QNI today remains principally in the descendants of the Oakley and Lindsay families who merged two local newspapers in 1926.

QNI entered the broadcast business in 1947, by signing on WQDI-FM (now WGEM-FM) in Quincy, the first FM station to serve the tri-state area. It purchased WGEM(AM) in 1948. In 1953, the company signed on the first television station in Quincy, WGEM-TV, an NBC affiliate.

In the early 1960s, the company partnered with Continental Cablevision on five cable systems in Illinois and Iowa. These systems were some of the first started by

---

[1] QNI takes its responsibility to serve the public interest very seriously. In addition to providing comprehensive news and public affairs programs, the Company supports its communities through numerous efforts both on and off the air. QNI's television stations, during 2000, provided over 65,000 no-charge public service announcements at a value of over $5 million and helped area groups raise over $20 million.

Continental. QNI sold its cable interest to Continental in 1974 when it began the acquisition of the additional five television stations which it now holds.

The six television stations operated by QNI include five NBC affiliates: KTIV Sioux City, Iowa, Market 144; KTTC Rochester, Minnesota, Market 153 WGEM-TV Quincy, Illinois, Market 161 WREX-TV Rockford, Illinois, Market 135; and WVVA Bluefield, West Virginia, Market 148.

The sixth station, WSJV, Elkhart/South Bend, Indiana, Market 87, is a Fox affiliate. QNI is currently in the process of acquiring five more television stations in Wisconsin to further expand its Midwest television cluster; these stations are all ABC affiliates.

B. QNI Has Been at the Forefront of New Technologies

QNI has been a leader in the early adoption of broadcast technology. In 1947, when QNI began operations of WQDI-FM in Quincy, Illinois, only a handful of FM's were on the air across the entire country. FM service didn't begin to penetrate and gain market share until the early 1970's. Another QNI station, WGEM(AM) was one of the first to broadcast in AM Stereo in 1976. WGEM-TV was an early adopter of color in 1962. Television stations KTTC, WGEM, WSJV and WVVA installed and began broadcasting in stereo in 1986 shortly after the BTSC standard was approved. We were one of the first broadcasters to use server-based technology for radio and later television. We have encouraged our local broadcast vendors to use our facilities as a test-bed for AM and FM IBOC (In band on channel) digital radio.

C. QNI's DTV Efforts

We have taken this same approach with respect to DTV. QNI has diligently planned for the DTV transition since the Telecommunications Act of 1996 became law. We attended all the early regional channel allocation meetings. We were also involved in submitting comments in the FCC's rulemaking process.

As early as the first quarter of 1997, QNI began choosing antennas, feed lines and commissioning tower studies to evaluate which of its towers would need to be replaced, and which ones could handle additional load. We knew early on that several of our towers weren't capable of or cost-effective to be used to support the addition of new antennas for digital. In August 1997, we formed a strategic partnership with Pappas Communications and Waitt Media to construct a new 2000-foot community tower in Sioux City, Iowa to serve the digital needs of the market, extend the service of existing stations, and add new television services to the market. Construction on this tower was completed in late 1998. This tower stands ready to be used for digital service on QNI's station KTIV as well as two other television stations in the market.

In January of 2000 we decided to move up our DTV launch of two of our stations, KTTC, Rochester and WGEM, Quincy, which are both required to be on-air by May 2002. WGEM went on the air on June 25, 2000 at a low power of 2.5kw. KTTC went on the air at full power on September 22, 2000 at 324kw. Both stations went on the air under Special Temporary Authority from the FCC, which was specifically applied for by QNI because the FCC had not yet issued construction permits for either station. We choose to launch DTV operations early for WGEM, Quincy, because of WGEM's connection to the broadcast manufacturing community in Quincy. In this regard, Quincy, Illinois is home to both Harris Communications and Broadcast Electronics. QNI frequently partners with these organizations to test and implement cutting edge technology. Currently WGEM is operating on the air testing Harris' Master Plus HD switcher. We decided to move forward early in Rochester because KTTC's service area encompasses affluent Olmstead County, Minnesota.[2] Counties with high affluency ratings tend to be early technology adopters.

In mid-2000, we finalized our DTV roll-out for our remaining four stations and the upgrade of WGEM to full power. Our plan called for all stations to be DTV-operational by October 1, 2001, well ahead of the mandatory May 2002 date. This plan, however, has changed. Not only are we going to miss our self-imposed October 1, 2001 deadline, but we may not be able to make the May 2002 deadline for these stations.

D. DTV Challenges

The difficulties we face are substantial. Certainly, the confusing and ongoing debate with respect to a transmission standard (*8VSB* v. *COFDM*) muddied the water and has caused delays. Hopefully this issue is now resolved. But we have also expe-

[2] According to Nielsen Research, Olmstead County has one of the highest affluency ratings in the United States, with a median household income of $43,977. This places Olmstead County in the top 40 counties of per capita income for the entire United States.

rienced problems concerning confused consumer response to DTV, lack of promotion and understanding of DTV by electronics retailers, cable system carriage, FCC processing, equipment availability, lack of programming, and costly and unrecoverable expenditures. We briefly highlight some of these problems below.

As to consumer response, after extensive promotion of the new digital service QNI has offered in the Rochester and Quincy markets, we are aware of only three high definition television sets in use in Rochester and five in Quincy, although well over 600 digital-ready sets have been sold in these markets. This is likely due in part to the fact that HDTV receivers and converters have not been readily available for purchase at area electronics stores, or if they are available, are not being promoted or even displayed properly. Our General Manager of KTTC visited one such store in Rochester that did have a DTV tuner but it was not demonstrating digital programming. The store manager and staff simply didn't understand, and don't really care to understand, what broadcast DTV is all about. Major chain stores and local consumer electronics stores have large inventories and don't want to lose a sale by confusing a customer with this new digital product.

In the area of cable carriage, with our channel allotment for Rochester, we launched a digital UHF service in a predominantly VHF market, with 82 percent cable penetration. However, it has been our experience that cable systems aren't interested in discussing retransmission or must carry of a digital signal.

With respect to FCC construction permits, the FCC is processing a substantial number of DTV applications. As a result, the majority of our stations' DTV construction permits were recently issued in January 2001, yet are still subject to the May 2002 deadline. One CP is currently being resubmitted to the FCC to resolve Canadian border issues—issues that weren't clarified until the fall of 2000. Another permit contains an error which will have to be corrected, and will thereby result in additional processing delays. Signing purchase contracts for millions of dollars for antennas, transmission line, transmitters, and towers without granted construction permits presented unreasonable financial risk.

Equipment delays have not become a serious problem for us yet, but they will be. We have been told by manufacturers that lead-time for transmitter orders is 4 to 6 months, for antenna orders is 7 to 10 months, and for guy wires is 3 months. This assumes, of course, that we have existing towers that can handle the additional load. If not, tower construction is a 2-year project.

The biggest delays even now are for installation service for antennas and transmission line on towers. The impact on the transition of a scarcity of tower construction crews should not be underestimated. The ramifications of the demand on and shortage of tower crews are far-reaching. Before the transition began, it was not uncommon to find same-day service to fix a transmission line burnout for a television or FM radio station. Today we are told that it may be several weeks or up to a month before a crew becomes available to make the repair. In the interim, a station in all likelihood may find itself off the air during this time or operating at very reduced power.[3]

It must also be recognized that this conversion to DTV is not a financial boon for broadcasters. From the very beginning it became clear that no definitive business plan exists for the DTV conversion. To try to anticipate avenues which would allow QNI to recoup a portion of what we are spending for DTV, we have considered the future use of part of the spectrum for data service. To date, QNI hasn't made any agreement to utilize our spectrum space for data transmission. From our standpoint, very little "upside" or benefit exists in this arena to offset the major capital investment we plan to make to remain in the television business. At this point, multicasting may offer a more realistic and immediate upside. Multicasting would allow us to expand services to the public and use the spectrum as a distribution tool for new services.

E. Meeting the Challenges

In spite of the numerous obstacles we face, in 2001 we will proceed to make necessary modifications to buildings, power systems, and microwave and fiber interconnections to support DTV implementation. We will install transmitters and mount remaining antenna and line in 2002. QNI has already spent $4.5 million on the transition and expects to spend another $12-$14 million. We have recently started construction on a new digital operations center in Quincy, Illinois to help manage

---

[3] In reference to zoning, QNI has been fortunate and hasn't faced major local zoning or permit issues. However, in many areas of the country, broadcast efforts have come to a complete stop due to a lack of local zoning approval. The transition is further complicated for stations in the northern part of the United States where it is not uncommon to only have five or 6 months of the year during which tower crews can install or erect systems due to weather conditions.

personnel, traffic, and billing and to help operate the stations more efficiently and ease the financial burden of the DTV rollout. We are dedicated to the DTV transition and are farther along in the process than many broadcasters are at this point, especially in the smaller markets. But we recognize that substantial challenges remain.

In meeting the challenges of the transition, broadcasters, particularly those in medium and small markets, need assistance—not additional obstacles. We respectfully submit that broadcasters must continue to be granted flexibility in creating digital services so that digital spectrum can be used for multicasting and data services in addition to HDTV. QNI also urges flexibility in the rollout schedule rather than rigid adherence to the construction deadline of May 2002, which is simply not realistic. We also strongly support the NAB and other broadcasters who have called for the following: 1. Interoperability standards for DTV and cable products; 2. Must-carry, including analog and digital must-carry during the transition; 3. DTV receiver performance standards; and 4. DTV reception capability in every TV receiver.

Without this support, it doesn't matter how aggressively we approach the transition or how we promote the potential for DTV. Without support, we will be unable as broadcasters to make a successful transition to digital television.

Prepared Statement of Robert T. Miller, President, Viacel Corp.

Viacel was asked to be a witness at the House Subcommittee on Telecommunications last summer and was the designated "datacaster" at that hearing. We feel datacasting is a valid use of the DTV spectrum and allows that spectrum to be used very efficiently. We further understand that any hearing cannot possibly cover all aspects of such a complicated subject as broadcasting in the digital age. We therefore have asked some questions of the broadcasters and proposed some answers here that were not asked in the hearings.

Of the broadcasters represented by Mr. Jeff Sagansky of Paxson Communications and Mr. Ben Tucker of Fisher Broadcasting I would ask: during most of your testimony you both strongly suggested that broadcasters needed "must-carry." In fact, in one exchange between Mr. Sagansky and one of the Senators a joking reference to "going in the cellar and getting the old rabbit ears out or putting an antenna up on the roof" was made, the implication being that this was a horrible outcome if "must-carry" was not extended to all proposed broadcast content. You suggested that 70 percent of your viewers received their TV through the marvels of cable and another 15 percent used satellite reception for a total of 85 percent. And, as we know, these numbers are growing. Soon they will be 90 percent and then 95 percent.

Your strong argument for "must-carry" seems to suggest that you do not now, or soon will not, need any spectrum at all. Other than a declining over-the-air viewership of 15 percent, you only use your spectrum to get your signal to the cable headend. You don't need it at all to get to the satellite headend. That feed to the cable headend could be replaced with a T-3 line or a piece of fiber and you may have already done that.

I believe Mr. Bud Paxson made a statement recently to the effect that he would be willing to give up channels in the 60 through 69 range in exchange for money and "must-carry," in effect leaving the ranks of broadcasters and becoming a content provider to cable companies, protected by Government decree. Your testimony suggests that other broadcasters are thinking the same way.

In other words, "must-carry" is all-important and actual broadcasting is a joke about "rabbit ears." You don't actually expect many people to receive over-the-air broadcast. In fact the digital TV 8-VSB modulation standard you recently voted to retain doesn't even reach as many people as the current NTSC analog system, does it? In fact, it would be receivable by just 92 percent of those 15 percent still dependent on over-the-air reception of TV, according to your recent NAB/MSTV tests.

Who are those 15 percent whose NTSC analog TV service will be turned off? Probably they are people who cannot afford cable and satellite. They surely are not the early adopters who are buying HDTV sets. These people will lose all access to free over-the-air TV, won't they?

And who will the free over-the-air DTV viewers be? Well-heeled suburbanites who can afford cable and satellite and HDTV sets and 8-VSB receivers. In fact, you may find that most of them will have all three.

Maintaining the fiction that you are broadcasters is very important to you. You tread a fine line between stating that almost everyone receives your signal over cable and admitting that you wouldn't mind turning off your transmitters and saving the power bills, because if that veil falls your legitimacy to demand "must-carry" of your content evaporates also. That veil is dropping daily as fewer people rely on over-the-air reception of your signal and sign up for cable and satellite. The incredible under-use of this "beachfront property" spectrum is becoming more and more apparent.

The "emperor has no clothes" is translating to "the broadcaster has no-over-the-air viewers." Nationally, there are 1600 NTSC analog stations plus 1600 ATSC digital stations, each with 6 MHz of spectrum, which comes to 38,400 MHz of prime spectrum being used to reach fewer and fewer viewers.

4435 MHz of "C&F BLOCK" PCS not so prime cellular spectrum just went at auction for $16,857,046,150.00 or $3,800,912.32 per MHz in January 2001. Now if it is comparable to the 38,400 MHz of TV spectrum lent to the broadcaster (maybe the TV spectrum is more valuable), then 38,400 MHz of TV spectrum would be worth $145,955,033,088.00, or roughly $146 billion.

As the number of viewers steadily drops, as more and more of them move to cable and satellite where our broadcaster friends tell us their signals must be carried, the rationale for using this national $146 billion resource declines as well. Nowhere in their testimony did they talk about more people receiving over-the-air free DTV. They talked about forcing every TV purchaser to buy a potentially expensive DTV tuner but then they laugh in mock terror at the notion that it may "come to" going in the basement to find the old rabbit ears or, heaven forbid, going up on the roof and strapping a Yagi antenna to the chimney.

Their tests, the NAB/MSTV tests, showed that with 8-VSB modulation, the one they recently settled on, 30 ft. antennas are the only way to receive free over-the-air DTV in the United States. Even then the tests showed only a 70 percent plus success rate, and probably a less than 20 percent success rate in the typical city, which they didn't even bother to test. The rabbit ear antenna which most of us would probably actually be able to use had a failure rate of 70 percent plus. So you won't and I won't and most of us are not going to be getting our HDTV over-the-air free, now are we? We are going to wait for cable and satellite. That is what the sales numbers say. Of 384,000 HDTV-ready monitors sold, only between 7,000 and 70,000 were sold with 8-VSB tuners for over-the-air reception. This means that those folks who could afford to spend $3,000.00 to $25,000.00 on HDTV monitors and everything that goes with it, would not part with a measly $500.00 more for a 8-VSB receiver. That is pretty telling. Sounds like with all the problems with over-the-air reception people just want to wait for cable and satellite.

Well, which is it? Are we all going to enjoy free over-the-air TV or are we all going to get our HDTV from cable and satellite? From the testimony of the broadcasters, it appears it will be overwhelmingly cable and satellite.

So they don't need any spectrum at all except for that nagging problem of the poor 15 percent that can't afford cable or satellite. That number will decline, it is assumed, by the transition date of 2006 or 2015, or 3000 if you are a pessimist, to something less than 15 percent, lets say 5 percent by 2015. We could reduce the spectrum allocated to broadcasters to 1 or 2 MHz instead of 6 MHz and let them continue broadcasting NTSC free over-the-air and have them deliver converter equipment so that the viewer could receive the digital signal and convert it to NTSC on their current analog sets.

This would free up 35,200 MHz at least of the 38,400 MHz that the broadcasters are now using. The cost of converter boxes for the 5 percent still relying on over-the-air reception could be paid for out of proceeds from the auction of the 35,200 MHz. Those proceeds would exceed the nominal $134 billion suggested above for no other reason than inflation till the year 2015. If a converter box costs $300 and 5 percent of 100 million TV homes needed one, this would come to 5 million times $300.00, which is $1.5 billion, or just over 1 percent of the $134 billion received.

So we give the broadcasters what they want, which is "must-carry" on cable and satellite. They can remain legitimate broadcasters with 2 MHz of spectrum broadcasting to 5 percent of the population (no one loses free over-the-air TV), the Treasury gets a windfall of $134 billion, everyone gets as much HDTV as they want due to the wonders of the free market, and everyone is happy.

The Modulation Debate

The other question of both the Senators and the broadcasters is why does Congress and its operative the FCC not exercise more direct control of the process of picking the modulation method that the United States citizens will suffer with for who knows how many years? The original picking and the subsequent retesting of the mandated 8-VSB and its rival the COFDM standards was left up to biased industry groups such as the NAB and the MSTV. The FCC then relies on these tests to issue orders that "reaffirm" that 8-VSB is indeed the modulation for the US.

The very fact that the tiresome phrase "the modulation debate has been decided, its over" is repeated so emphatically and so often by the biased parties tends to confirm what these hearings and further hearings in the House also tend to confirm, it is not over until we get it right and the digital TV transition is accomplished.

Saying that the modulation debate is over is hike treating a patient in a trauma emergency ward for superficial cuts and bruises while ignoring the fact that he is not breathing. You can talk all you want about the other big problems of the digital transition but it will remain dead until you deal with the central issue. 8-VSB is not working and we have no idea when it might be made to work in the future.

Congress should take a good long look at the RFP that the ATSC published to "fix" 8-VSB. It is a specification sheet for COFDM. In other words COFDM already does everything we want to fix about 8-VSB. If 8-VSB could and were indeed fixed according to the dictates of this RFP it would then be a closer relative to COFDM then to the current 8-VSB. It would also quite possibly be a new standard that is incompatible with current receivers and have to be reviewed by the FCC all over again. We would be, in a few years if we are lucky with "inventions" that fix 8-VSB, in the same boat that we are in now with COFDM. Why should we wait? Why not allow COFDM now and see what happens to 8-VSB in the inventor's hands. Current 8-VSB receivers would go on working, maybe only a few broadcasters would even attempt COFDM since the vote on January 15, 2001 was 27 for 8-VSB and only 3 for COFDM. It wouldn't be any fuss since the MSTV/NAB test showed that there was no difference between the interference patterns of COFDM and 8-VSB.

Why do I bring up this debate again? Because if a working modulation standard were allowed then companies like ours could hatch business plans that actually might accomplish the digital transition without all the fuss.

Our plan calls for a free service with the distribution of free receivers to millions of viewers that would not impact the broadcast of HDTV or SDTV. We need the broadcast spectrum to actually reach people over-the-air. We don't need "must-carry" we need "must-receive."

We believe our plan will have copycats as soon as we start. The transition could be over before you know it.

The Conspiracy Theory

If we believe that the broadcasters or most of them want must-carry at almost any cost then we can imagine some perverted reasoning going on in their content centric heads. For example, the cable and satellite industries both were initially invented to solve TV's biggest problem, reception. Both NTSC and the new digital ATSC both have similar reception problems. In fact the new digital ATSC standard, 8-VSB, is slightly worse in reception than the old NTSC (ATSC equals 92 percent of NTSC according to the MSTV test). The very reason for "must-carry" is related to the reception problems of NTSC. Jeff and Ben strongly argued for "must-carry" because otherwise their viewers could "not receive" their content. That seems to be their main argument for "must-carry."

If broadcasters who are already having a major problem pushing through "must-carry" even with their reception affliction suddenly were cured that could cause them a problem.

If they as broadcasters have no problem with reception by their customers in their coverage area what would be the rationale for "must-carry"? They have already argued that all TVs be sold with digital receivers so if there was no problem with reception and everyone has receivers they have no need for "must-carry" right?

COFDM does solve the problem of reception. It is the biggest nightmare of the big broadcasters who depend on content delivery to cable companies. They will do and have done anything to muddy the waters about COFDM's capabilities. They have attacked every test around the world that showed COFDM to be far superior. They went to great lengths to make sure that COFDM did not get a fair treatment in the MSTV test.

First they tried to force COFDM to operate at a lower power, and then they operated in secrecy and with precision zeroed in on one piece of equipment that could be called a professional COFDM receiver only if you translate the English phrase "transmitter monitor" through a random series of foreign tongues till by chance the term "receiver" pops out.

In Conclusion

The problem with the digital transition is twofold. The digital transition has been kidnapped by the HDTV virus. First we should concentrate on the DIV transition.

Second the political process has failed us. The regulatory agency, the FCC has been kidnapped by the industries they regulate. Biased and special interest have intruded into the decisionmaking process and an unscientific decision had been made as to the technology we should use in our digital transition.

COFDM is one of the correct answers, 8-VSB is a wrong answer.

Companies like Viacel Corporation would flood the market with free digital TV receivers and non-subscription free services quickly if a working and receivable digital modulation standard were adopted, (just allowed is good enough), COFDM receivers are already available for as little as $100.00 wholesale.

○

APEX
TP
9780160744822
76901953 – 25